TENTH EDITION

THE ORGANIC CHEM LAB SURVIVAL MANUAL

A Student's Guide to Techniques

JAMES W. ZUBRICK
Hudson Valley Community College

WILEY

Vice President & Director	Petra Recter
Senior Acquisitions Editor	Nick Ferrari
Sponsoring Editor	Joan Kalkut
Product Designer	Sean Hickey
Project Manager	Gladys Soto
Project Specialist	Marcus Van Harpen
Marketing Solutions Assistant	Mallory Fryc
Senior Marketing Manager	Kristine Ruff
Assistant Marketing Manager	Puja Katariwala
Associate Director	Kevin Holm
Photo Editor	Nicholas Olin
Production Editor	Ameer Basha

This book was set in 10/12 Times by SPi Global and printed and bound by Donnelley/Harrisonburg.

Founded in 1807, John Wiley & Sons, Inc. has been a valued source of knowledge and understanding for more than 200 years, helping people around the world meet their needs and fulfill their aspirations. Our company is built on a foundation of principles that include responsibility to the communities we serve and where we live and work. In 2008, we launched a Corporate Citizenship Initiative, a global effort to address the environmental, social, economic, and ethical challenges we face in our business. Among the issues we are addressing are carbon impact, paper specifications and procurement, ethical conduct within our business and among our vendors, and community and charitable support. For more information, please visit our website: *www.wiley.com/go/citizenship*.

Evaluation copies are provided to qualified academics and professionals for review purposes only, for use in their courses during the next academic year. These copies are licensed and may not be sold or transferred to a third party. Upon completion of the review period, please return the evaluation copy to Wiley. Return instructions and a free of charge return mailing label are available at *www.wiley.com/go/returnlabel*. If you have chosen to adopt this textbook for use in your course, please accept this book as your complimentary desk copy. Outside of the United States, please contact your local sales representative.

ISBN: 978-1-118-87578-0 (PBK)

Library of Congress Cataloging-in-Publication Data

Names: Zubrick, James W.
Title: The organic chem lab survival manual : a student's guide to techniques
/ James W. Zubrick, Hudson Valley Community College.
Description: 10th edition. I Hoboken, NJ : John Wiley & Sons, Inc., 2015. I
Includes bibliographical references and index.
Identifiers: LCCN 2015021960 I ISBN 9781118875780 (pbk.)
Subjects: LCSH: Chemistry, Organic—Laboratory manuals.
Classification: LCC QD261 .Z83 2015 I DDC 547.0078—dc23 LC record available at
http://lccn.loc.gov/2015021960

Printing identification and country of origin will either be included on this page and/or the end of the book. In addition, if the ISBN on this page and the back cover do not match, the ISBN on the back cover should be considered the correct ISBN.

Printed in the United States of America

10 9 8 7 6 5 4 3 2 1

PREFACE TO THE TENTH EDITION

This, the tenth edition, brings changes from circumstances both imposed and self-prompted, while retaining the emphasis on the basic techniques and doing the work correctly the first time. I have relied on the comments of users and reviewers to point out changes that have been adopted and my own observations of my own students in the laboratory and elsewhere to introduce other modifications.

Earlier material on "The Theory of Distillation" (Chapter 34) and the detailed description of the operation of a dual-beam IR (Chapter 32, "Infrared Spectroscopy") have been moved online at Wiley www.wiley.com/college/zubrick reflecting current teaching trends pointed out by reviewers. The entirely separate section on "Theory of Extraction" has been moved into Chapter 15 ("Extraction and Washing") to provide a tighter presentation, and changes have been made to the text and artwork in that chapter for a more modern appearance.

The artwork for "Keeping a Notebook" (Chapter 2) has been redrawn, reflecting the change in notebook pages to a gridded "engineering format," although the two basic experiments, a technique and a synthesis, have remained unchanged. A new section on writing notebook entries "by the numbers," with examples and reasons why we write what we do, when we do, may not eliminate the concern often approaching panic students have about "what should I write," but I introduce both "The Golden Rule" and "The Silver Rule" in an effort to keep research notebook keepers on the straight and narrow.

"The Melting Point Experiment" (Chapter 12) now includes the SRS DigiMelt, a 21st century digital version of the Mel-Temp, joining the previous, more classical instrumentation and techniques.

The biggest change, and for me, personally, the biggest disappointment, is having to acknowledge and adapt to the fact that chemistry handbooks have been supplanted by Android phones and an Internet link. It's been ten years since a new edition of Lange's "Handbook of Chemistry" has been published; more's the pity as Norbert Lang was a co-editor of the other chemistry bible, The CRC Handbook of Chemistry and Physics, when he struck out to start his own in the middle of The Great Depression, possibly because there was starting to be more physics than chemistry in The CRC Handbook. Nonetheless, we soldier on to help people navigate the new waters, pointing out the sights and shoals.

It takes a lot of effort from a lot of people to produce such a work. I'd like to thank my reviewers, Lucy Moses, Virginia Commonwealth University; Christine Rich, University of Louisville; Sean O'Connor, University of New Orleans;

Jeffrey Hugdahl, Mercer University; Kathleen Peterson, University of Notre Dame; Chavonda Mills, Georgia College & State University; Beatrix Aukszi, Nova Southeastern University; Robert Stockland, Bucknell University; Jennifer Krumper, UNC-Chapel Hill; Rui Zhang, Western Kentucky University; Holly Sebahar, University of Utah; Adam List, Vanderbilt University for their comments and suggestions, most of which have been incorporated in this work. Finally, I'd like to thank Petra Recter, Associate Publisher, Chemistry and Physics, for the chance to perform this update, and Joan Kalkut, Sponsoring Editor, for her tremendous patience and support during a personally difficult time.

J. W. Zubrick
Hudson Valley Community College

CONTENTS

SAFETY FIRST, LAST, AND ALWAYS

- *Wear your goggles over your eyes.*
- *If you don't know where a waste product goes—ASK!*
- *Careful reading can prevent failure.*

The organic chemistry laboratory is potentially one of the most dangerous of undergraduate laboratories. That is why you must have a set of safety guidelines. It is a very good idea to pay close attention to these rules, for one very good reason:

The penalties are only too real.

Disobeying safety rules is not at all like flouting many other rules. *You can get seriously hurt.* No appeal. No bargaining for another 12 points so you can get into medical school. Perhaps as a patient, but certainly not as a student. So, go ahead. Ignore these guidelines. But remember—

You have been warned!

1. *Wear your goggles.* Eye injuries are extremely serious but can be mitigated or eliminated if you keep your goggles on *at all times*. And I mean *over your eyes,* not on top of your head or around your neck. There are several types of eye protection available, some of them acceptable, some not, according to local, state, and federal laws. I like the clear plastic goggles that leave an unbroken red line on your face when you remove them. Sure, they fog up a bit, but the protection is superb. Also, think about getting chemicals or chemical fumes trapped under your contact lenses before you wear them to lab. Then don't wear them to lab. Ever.

2. ***Touch not thyself.*** Not a Biblical injunction, but a bit of advice. You may have just gotten chemicals on your hands in a concentration that is not noticeable, and, sure enough, up go the goggles for an eye wipe with the fingers. Enough said.

3. ***There is no "away."*** Getting rid of chemicals is a very big problem. You throw them out from here, and they wind up poisoning someone else. Now there are some laws to stop that from happening. The rules were really designed for industrial waste, where there are hundreds of gallons of waste that all has the same composition. In a semester of organic lab, there will be much smaller amounts of different materials. Waste containers could be provided for everything, but this is not practical. If you don't see the waste can you need, ask your instructor. When in doubt, *ask.*

4. ***Bring a friend.*** *You must never work alone.* If you have a serious accident and you are all by yourself, you might not be able to get help before you die. Don't work alone, and don't work at unauthorized times.

5. ***Don't fool around.*** Chemistry is serious business. Don't be careless or clown around in lab. You can hurt yourself or other people. You don't have to be somber about it—just serious.

6. ***Drive defensively.*** Work in the lab as if someone else were going to have an accident that might affect you. Keep the goggles on because *someone else* is going to point a loaded, boiling test tube at you. *Someone else* is going to spill hot, concentrated acid on your body. Get the idea?

7. ***Eating, drinking, or smoking in lab.*** Are you kidding? Eat in a chem lab?? Drink in a chem lab??? Smoke, and blow yourself up????

8. ***The iceman stayeth, alone.*** No food in the ice machine. "It's in a plastic bag, and besides, nobody's spilled their product onto the ice yet." No products cooling in the ice machine, all ready to tip over, either. Use the scoop, and nothing but the scoop, to take ice out of the machine. And don't put the scoop in the machine for storage, either.

9. ***Keep it clean.*** Work neatly. You don't have to make a fetish out of it, but try to be *neat.* Clean up spills. Turn off burners or water or electrical equipment when you're through with them. Close all chemical containers after you use them. Don't leave a mess for someone else.

10. ***Where it's at.*** Learn the locations and proper use of the fire extinguishers, fire blankets, safety showers, and eyewash stations.

11. ***Making the best-dressed list.*** Keep yourself covered from the neck to the toes—no matter what the weather. That might include long-sleeved tops that also cover the midsection. Is that too uncomfortable for you? How about a chemical burn to accompany your belly button, or an oddly shaped scar on your arm in lieu of a tattoo? Pants that come down to the shoes and cover any exposed ankles are probably a good idea as well. No open-toed shoes, sandals, or canvas-covered footwear. No loose-fitting cuffs on the pants or the shirts. Nor are dresses appropriate for lab. Keep the midsection covered. Tie back

that long hair. And a small investment in a lab coat can pay off, projecting that extra professional touch. It gives a lot of protection, too. Consider wearing disposable gloves. Clear polyethylene ones are inexpensive, but the smooth plastic is slippery, and there's a tendency for the seams to rip open when you least expect it. Latex examination gloves keep their grip and don't have seams, but they cost more. Gloves are not perfect protectors. Reagents like bromine can get through and cause severe burns. They'll buy you some time, though, and can help mitigate or prevent severe burns. Oh, yes—laboratory aprons: They only cover the *front,* so your exposed legs are still at risk from behind.

12. ***Hot under the collar.*** Many times you'll be asked or told to heat something. Don't just automatically go for the Bunsen burner. That way lies *fire.* Usually—

No flames!

Try a hot plate, try a heating mantle (see Chapter 17, "Sources of Heat"), but try to stay away from flames. Most of the fires I've had to put out started when some bozo decided to heat some flammable solvent in an open beaker. Sure, there are times when you'll *have* to use a flame, but use it away from all flammables and in a hood (Fig. 1.1), and only with the permission of your instructor.

13. ***Work in the hood.*** A hood is a specially constructed workplace that has, at the least, a powered vent to suck noxious fumes outside. There's also a safety glass or plastic panel you can pull down as protection from exploding apparatus (Fig. 1.1). If it is at all possible, treat every chemical (even solids) as if toxic or bad-smelling fumes can come from it, and carry out as many of the operations in the organic lab as you can *inside a hood,* unless told otherwise.

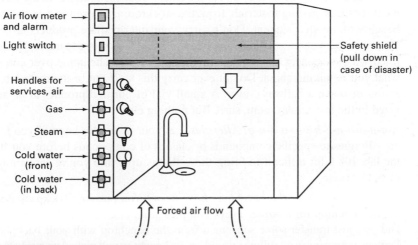

FIGURE 1.1 A typical hood.

14. ***Keep your fingers to yourself.*** Ever practiced "finger chemistry"? You're unprepared so you have a lab book out, and your finger points to the start of a sentence. You move your finger to the end of the first line and do that operation—

    "Add this solution to the beaker containing the ice-water mixture"

    And WHOOSH! Clouds of smoke. What happened? The next line reads—

    "very carefully as the reaction is highly exothermic."

    But you didn't read that line, or the next, or the next. So you are a danger to yourself and everyone else. Read and take notes on any experiment before you come to the lab (see Chapter 2, "Keeping a Notebook").

15. ***Let your eyes roam.*** Not over to another person's exam paper, but all over the entire label of any reagent bottle. You might have both calcium carbonate and calcium chloride in the laboratory, and if your eyes stop reading after the word "calcium," you have a good chance of picking up and using the wrong reagent. At the very least, your experiment fails quietly. You don't really want to have a more exciting exothermic outcome. Read the entire label and be sure you've got the right stuff.

16. ***What you don't know can hurt you.*** If you are not sure about an operation, or you have any question about handling anything, *please* ask your instructor before you go on. Get rid of the notion that asking questions will make you look foolish. Following this safety rule may be the most difficult of all. Grow up. Be responsible for yourself and your own education.

17. ***Blue Cross or Blue Shield?*** Find out how you can get medical help if you need it. Sometimes, during a summer session, the school infirmary is closed, and you would have to be transported to the nearest hospital.

18. ***What's made in Vegas, stays in Vegas.*** You're preparing a compound, and you have a question about what to do next. Perhaps your instructor is in the instrument room, or getting materials from the stockroom, or even just at the next bench with another student. Don't carry your intermediate products around; go *a capella* (without accompaniment of beakers, flasks, or separatory funnels filled with substances) to your instructor and ask that she come over and see what you're talking about. Do not ever carry this stuff out of the main lab, or across or down a hallway—ever. A small vial of purified product to be analyzed in the instrument room, sure. But nothing else.

19. ***A-a-a-a-a-a-c-h-o-o-o-o-o-o! Allergies.*** Let your instructor know if you have any allergies to specific compounds or classes of compounds before you start the lab. It's a bit difficult to bring these things up while you're scratching a rash. Or worse.

20. ***Do you know where the benchtops have been?*** You put your backpack down on the benchtop for a while. Then, you pick it up and put it somewhere else. Did you just transfer some substance from the benchtop with your backpack? Perhaps your pens were rolling around on the benchtop and picked up a substance

themselves and you didn't know it? Often wearing protection doesn't help; gloves can transfer chemicals to your pen (and you can't tell because your hands are covered), and that pen might go where? Behind the ear? In the mouth?

These are a few of the safety guidelines for an organic chemistry laboratory. You may have others particular to your own situation.

ACCIDENTS WILL NOT HAPPEN

That's an attitude you might hold while working in the laboratory. You are *not* going to do anything or get anything done to you that will require medical attention. If you do get cut, and the cut is not serious, wash the area with water. If there's serious bleeding, apply direct pressure with a clean, preferably sterile, dressing. For a minor burn, let cold water run over the burned area. For chemical burns to the eyes or skin, flush the area with lots of water. In every case, get to a physician if at all possible.

If you have an accident, *tell your instructor immediately. Get help!* This is no time to worry about your grade in lab. If you put grades ahead of your personal safety, be sure to see a psychiatrist after the internist finishes.

DISPOSING OF WASTE

Once you do your reaction, since your mother probably doesn't take organic lab with you, you'll have to clean up after yourself. I hesitated to write this section for a very long time because the rules for cleaning up vary greatly according to, but not limited to, federal, state, and local laws, as well as individual practices at individual colleges. There are even differences—legally—if you or your instructor do the cleaning up. And, as always, things do seem to run to money—the more money you have to spend, the more you can throw away. So there's not much point in even trying to be authoritative about waste disposal in this little manual, but there are a few things I have picked up that you should pay attention to. Remember, my classification scheme may not be the same as the one you'll be using. When in doubt, *ask! Don't just throw everything into the sink. Think.*

> *Note to the picky: The word nonhazardous, as applied here, means relatively benign, as far as organic laboratory chemicals go. After all, even pure water, carelessly handled, can kill you.*

How you handle laboratory waste will depend upon what it is. Here are some classifications you might find useful:

1. *Nonhazardous insoluble waste.* Paper, corks, sand, alumina, silica gel, sodium sulfate, magnesium sulfate, and so on can probably go into the ordinary wastebaskets in the lab. Unfortunately, these things can be contaminated with hazardous waste (see the following items), and then they need special handling.

2. *Nonhazardous soluble solid waste.* Some organics, such as benzoic acid, are relatively benign and can be dissolved with a lot of tap water and flushed down the drains. But if the solid is that benign, it might just as well go out with the nonhazardous insoluble solid waste, no? Check with your instructor; watch out for contamination with more hazardous materials.

3. *Nonhazardous soluble liquid waste.* Plain water can go down the drains, as well as water-soluble substances not otherwise covered below. Ethanol can probably be sent down the drains, but butanol? It's not that water soluble, so it probably should go into the general organic waste container. Check with your instructor; watch out for contamination with more hazardous materials.

4. *Nonhazardous insoluble liquid waste.* These are compounds such as 1-butanol (previously discussed), diethyl ether, and most other solvents and compounds not covered otherwise. In short, this is the traditional "organic waste" category.

5. *Generic hazardous waste.* This includes pretty much all else not listed separately. Hydrocarbon solvents (hexane, toluene), amines (aniline, triethylamine), amides, esters, acid chlorides, and on and on. Again, traditional "organic waste." Watch out for incompatibilities, though, before you throw just anything in any waste bucket. If the first substance in the waste bucket was acetyl chloride and the second is diethylamine (both hazardous liquid wastes), the reaction may be quite spectacular. You may have to use separate hazardous waste containers for these special circumstances.

6. *Halogenated organic compounds.* 1-Bromobutane and *tert*-butyl chloride, undergraduate laboratory favorites, should go into their own waste containers as "halogenated hydrocarbons." There's a lot of agreement on this procedure for these simple compounds. But what about your organic unknown, 4-bromobenzoic acid? I'd have you put it and any other organic with a halogen in the "halogenated hydrocarbon" container and not flush it down the drain as a harmless organic acid, as you might do with benzoic acid.

7. *Strong inorganic acids and bases.* Neutralize them, dilute them, and flush them down the sink. At least as of this writing.

8. *Oxidizing and reducing agents.* Reduce the oxidants and oxidize the reductants before disposal. Be careful! Such reactions can be highly exothermic. Check with your instructor before proceeding.

9. *Toxic heavy metals.* Convert to a more benign form, minimize the bulk, and put in a separate container. If you do a chromic acid oxidation, you might reduce the more hazardous C^{6+} to Cr^{3+} in solution and then precipitate the Cr^{3+} as the hydroxide, making lots of expensive-to-dispose-of chromium solution into a tiny amount of solid precipitate. There are some gray areas, though. Solid manganese dioxide waste from a permanganate oxidation should probably be considered a hazardous waste. It can be converted to a soluble Mn^{2+} form, but should Mn^{2+} go down the sewer system? I don't know the effect of Mn^{2+} (if any) on the environment. But do we want it out there?

Mixed Waste

Mixed waste has its own special problems and raises even more questions. Here are some examples:

1. *Preparation of acetaminophen (Tylenol): a multistep synthesis.* You've just recrystallized 4-nitroaniline on the way to acetaminophen, and washed and collected the product on your Buchner funnel. So you have about 30–40 mL of this really orange solution of 4-nitroaniline and by-products. The nitroaniline is very highly colored, the by-products probably more so, so there isn't really a lot of solid organic waste in this solution, not more than perhaps 100 mg or so. Does this go down the sink, or is it treated as organic waste? Remember, you have to package, label, and transport to a secure disposal facility what amounts to 99.9% perfectly safe water. Check with your instructor.

2. *Preparation of 1-bromobutane.* You've just finished the experiment and you're going to clean out your distillation apparatus. There is a residue of 1-bromobutane coating the three-way adapter, the thermometer, the inside of the condenser, and the adapter at the end. Do you wash the equipment in the sink and let this minuscule amount of a halogenated hydrocarbon go down the drain? Or do you rinse everything with a little acetone into yet another beaker and pour that residue into the "halogenated hydrocarbon" bucket, fully aware that most of the liquid is acetone and doesn't need special halide treatment? Check with your instructor.

3. *The isolation and purification of caffeine.* You've dried a methylene chloride extract of caffeine and are left with methylene chloride–saturated drying agent. Normally a nonhazardous solid waste, no? Yes. But where do you put this waste while the methylene chloride is on it? Some would have you put it in a bucket in a hood and let the methylene chloride evaporate into the atmosphere. Then the drying agent is nonhazardous solid waste. But you've merely transferred the problem somewhere else. Why not just put the whole mess in with the "halogenated hydrocarbons"? Usually, halogenated hydrocarbons go to a special incinerator equipped with traps to remove HCl or HBr produced by burning. Drying agents don't burn very well, and the cost of shipping the drying agent part of this waste is very high. What should you do? Again, ask your instructor.

In these cases, as in many other questionable situations, I tend to err on the side of caution and consider that the bulk of the waste has the attributes of its most hazardous component. This is, unfortunately, the most expensive way to look at the matter. In the absence of guidelines,

1. Don't make a lot of waste in the first place.
2. Make it as benign as possible. (Remember, though, that such reactions can be highly exothermic, so proceed with caution.)
3. Reduce the volume as much as possible.

Oh: Try to remember that sink drains can be tied together, and if you pour a sodium sulfide solution down one sink while someone else is diluting an acid in another sink, toxic, gagging, rotten-egg-smelling hydrogen sulfide can back up the drains in your entire lab, and maybe even the whole building.

MATERIAL SAFETY DATA SHEET (MSDS)

The MSDS for any substance is chock-full of information, including but not limited to the manufacturer, composition (for mixtures), permissible exposure limit (PEL), threshold limit value (TLV) boiling point, melting point, vapor pressure, flash point, and on and on and on. These data sheets are very complete, very thorough, and very irrelevant to working in the undergraduate organic chemistry laboratory Period.

Don't take my word for it. One outfit, Interactive Learning Paradigms Incorporated (http://www.ilpi.com/msds/faq/parta.html), clearly states: "An MSDS reflects the hazards of working with the material in an occupational fashion. For example, an MSDS for paint is not highly pertinent to someone who uses a can of paint once a year, but is extremely important to someone who does this in a confined space 40 hours a week."

And probably less pertinent, if that's even possible, to someone who will work with 1-bromobutane once in a lifetime.

So if you're teaching organic lab, that's one thing. If you're taking organic lab, well, stick to hazard data and references in the other handbooks and you'll be knowledgeable enough.

GREEN CHEMISTRY AND PLANNING AN ORGANIC SYNTHESIS

While it is always good to "reduce, reuse, recycle," unless you're developing new experiments you don't really have any control over these things. But if you have to plan an organic synthesis from the ground up, might as well do it right.

1. *Eschew the older literature!* 'Fraid so. Many places will initially steer you to *Organic Syntheses,* which runs from 1932 to the present, as the syntheses there have been checked and will work as advertised. Unfortunately, for the early work there, and in many other places, being green just wasn't even thought about. So be careful. A historical collection of techniques in a reference with a current copyright date can detail reactions that would not be considered green today.

2. *Teaching over research.* A better place to look is *The Journal of Chemical Education,* rather than the traditional research resources. While a large research

group at a large university can have the resources (read *money*) to have toxic materials disposed of properly, "one-man shops" at community colleges are under greater pressure to reduce the costs of waste disposal, and, while they may not be the ones to originally develop a greener method from the high-powered research lab, they certainly exploit it, often in an inspired fashion.

3. *Make what you want, but use what you make.* You'll have to decide on just how much product you'll need to synthesize. And it depends upon the scale of your apparatus.
   - **Microscale.** For a **solid product**, target at least 200 mg. This should be enough for a melting point, an IR, and an NMR, plus some to hand in to show you made it. If you have to, you can easily recover your product from the NMR solvent; IR might be too problematic to bother about. For a **liquid product**, besides the tests, there might be drying and distillation, so about 2 mL might be your target. Don't forget to use the density of the liquid to calculate the mass you'll need to use for your stoichiometric calculations.
   - **Miniscale.** About 5 g for a solid; about 10 mL of a liquid. Just guidelines, now. The consequences of losing product at any stage are greatly reduced. Doesn't mean you should be sloppy with your technique, though.

4. *Plan to lose.* Now that you know how much you're planning to make, assume you won't be making it in a perfect yield. For first-time-this-has-ever-been-done reactions, you might get 40%; if the reaction has been done before, and you have a published procedure with a posted yield, but *you've* never done this before, add a 10% penalty. Then calculate back to get the amount of starting materials you'll need based on this lower yield.

5. *Timing is everything.* Generally, the reaction times shouldn't be reduced. Paradoxically, if you have the time, you can take the time to find out by running the experiment over and over again using different reaction times to find the best time. If the published procedure uses half-molar quantities (large-scale equipment), and you rework this for microscale, you might reduce the reaction time since the smaller quantity will have lower thermal mass and not need to be heated for as long a time. Maybe.

6. *Use less-toxic materials.* Easy to say; a bit more difficult to do. Some suggestions in no particular order:
   - Do you even need a reaction solvent? Consider direct combination of reagents.
   - Can you replace chlorinated solvents, especially in extraction? You might consider diethyl ether or ethyl acetate.
   - Can you eliminate toxic metals? A shift from a chromium-based to a manganese-based oxidizer in a reaction may help. Organic catalysts can substitute for those based on heavy metals. That sort of thing.

AN iBAG FOR YOUR iTHING

The Survival Manual is available on a Kindle, and along with iPads, Androids, and Nooks, it looks like electronic hardware might be on your benchtop along with everything else. Warrantees aside, though, if somebody's cooling hose pops off and it gets soaked, you might be at quite a loss for quite a while. There is, however, a high-tech remedy: Ziploc bags.

Once inside a bag, the nasty elements of the laboratory can't get to your Kindle, but you can still use your fingers to manipulate the screen. We put 10 in. diagonal screen tablet computers in large Ziploc bags, and not only did they survive water spills and such, we could still write on the screen through the bag with the stylus. They were a bit slipperier than the tablet screen, and we had to stretch the plastic bag a bit to flatten it out, but they worked out.

EXERCISES

1. Make a rough sketch of your lab. Mark where the fire extinguishers, fire blanket, eye wash station, and other safety equipment are, as well as where you'll be working.
2. Why shouldn't you work in a laboratory by yourself?
3. Might there be any problems wearing contact lenses in the laboratory?
4. Biology laboratories often have stools. Why might this be foolish in the organic chemistry laboratory?
5. What the heck are the PEL, TLV, and flash point of substances?
6. Google the MSDS for 2-naphthol. Try to select one from Thermo Fisher and another from J. T. Baker/Mallinckrodt. Speculate as to why one says this compound will cause death on inhalation, and the other, well, not so much. Google the MSDS for sugar, also.

KEEPING A NOTEBOOK

- *Take notes before lab; make notes during lab.*
- *Take the notebook to the balance.*
- *No blank spaces for "future values"; no backfilling.*

A **research notebook** is one of the most valuable pieces of equipment you can own. With it, you can duplicate your work, find out what happened at your leisure, and even figure out where you blew it. General guidelines for a notebook are as follows:

1. The notebook must be permanently bound. No loose-leaf or even spiral-bound note-books will do. It should have a sewn binding so that the only way pages can come out is to cut them out. ($8\frac{1}{2} \times 11$ in. is preferred.) **Duplicate carbonless notebooks** are available that let you make removable copies that you can hand in. (Don't forget the **cardboard separator**—or you'll make lots of copies of your latest labwork when your writing goes through to subsequent pages.) And if the pages aren't already numbered, you should do it yourself.

2. *Use waterproof ink! Never pencil!* Pencil will disappear with time, and so will your grade. Cheap ink will wash away and carry your grades down the drain. Never erase! Just draw *one* line through ~~yuor errers~~ your errors so that they can still be seen. And never, never, never cut any pages out of the notebook!

3. Leave a few pages at the front for a table of contents. These entries will probably be the titles of the experiments you perform. If you have one of those fancy-dancy copies of "The Official Laboratory Research Notebook," and on the inside cover they've "helpfully" printed a "Record of Contents": Two columns of 22 lines that are at best three inches long. Not nearly long enough.

If you look at either the "A Technique Experiment" or "A Synthesis Experiment" sections ahead, the first note in each is to use a descriptive title. Sure, the word "Distillation" will fit on that 3-inch line, but that doesn't tell anyone enough in a table of contents to make a decision about reading it or not.

Leave a few pages at the front for a table of contents.

4. Your notebook is your friend, your confidant. Tell it:
   a. What you have done. Not what it says to do in the lab book. What you, yourself, have done.
   b. Any and *all* observations: color changes, temperature rises, explosions . . . anything that occurs. Any *reasonable* explanation of *why* whatever happened, happened.

5. Skipping pages is in *extremely* poor taste. It is NOT done!

6. List the IMPORTANT chemicals you'll use during each reaction. You should include USEFUL **physical properties**: the name of the compound, molecular formula, molecular weight, melting point, boiling point, density, and so on. You might have entries for the number of moles and notes on handling precautions. Useful information, remember.

Note the qualifier "useful." If you can't use any of the information given, do without it! You look things up *before* the lab so you can tell what's staring back out of the flask at you during the course of the reaction.

Your laboratory experiments can be classified as either of two major types: a technique experiment or a synthesis experiment. Each type requires different handling.

A TECHNIQUE EXPERIMENT

In a technique experiment, you get to practice a certain operation *before* you have to do it in the course of a synthesis. Distilling a mixture of two liquids to separate them is a typical technique experiment.

Read the following handwritten notebook pages with some care and attention to the *typeset* notes in the margin. A thousand words are worth a picture or so (Figs. 2.1–2.3).

Notebook Notes

1. Use a descriptive title for your experiment. "Distillation." This implies you've done *all* there is in the *entire* field of distillation. You haven't? Perhaps all you've done is "The Separation of a Liquid Mixture by Distillation." Hmmmmmm.

2. Writing that first sentence can be difficult. Try stating the obvious.

3. There are no large blank areas in your notebook. Draw sloping lines through them. Going back to enter observations after the experiment is over is *not professional*. Initial and date pages anytime you write anything in your notebook.

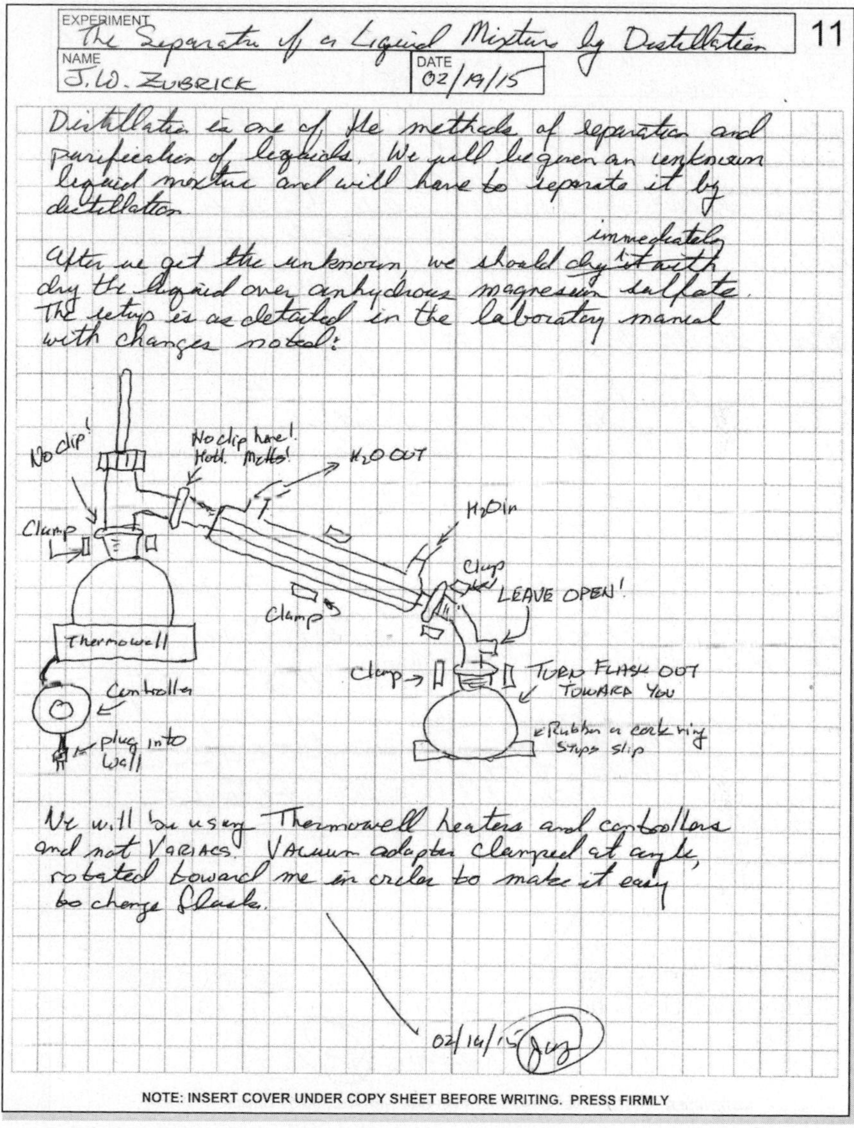
EXPERIMENT The Separation of a Liquid Mixture by Distillation 11

NAME J.W. ZUBRICK DATE 02/14/15

Distillation is one of the methods of separation and purification of liquids. We will be given an unknown liquid mixture and will have to separate it by distillation.

After we get the unknown, we should immediately dry it with dry the liquid over anhydrous magnesium sulfate. The setup is as detailed in the laboratory manual with changes noted:

We will be using Thermowell heaters and controllers and not VARIACS. Vacuum adapter clamped at angle, rotated toward me in order to make it easy to change flasks.

02/14/15 JWZ

NOTE: INSERT COVER UNDER COPY SHEET BEFORE WRITING. PRESS FIRMLY

FIGURE 2.1 Notebook entry for a technique experiment (1).

4. Note the appropriate changes in verb tense. Before you do the work, you might use the present or future tense when you write about something that *hasn't happened yet.* During the lab, since you are supposed to write what you've actually done just after you've actually done it, a simple past tense is sufficient.

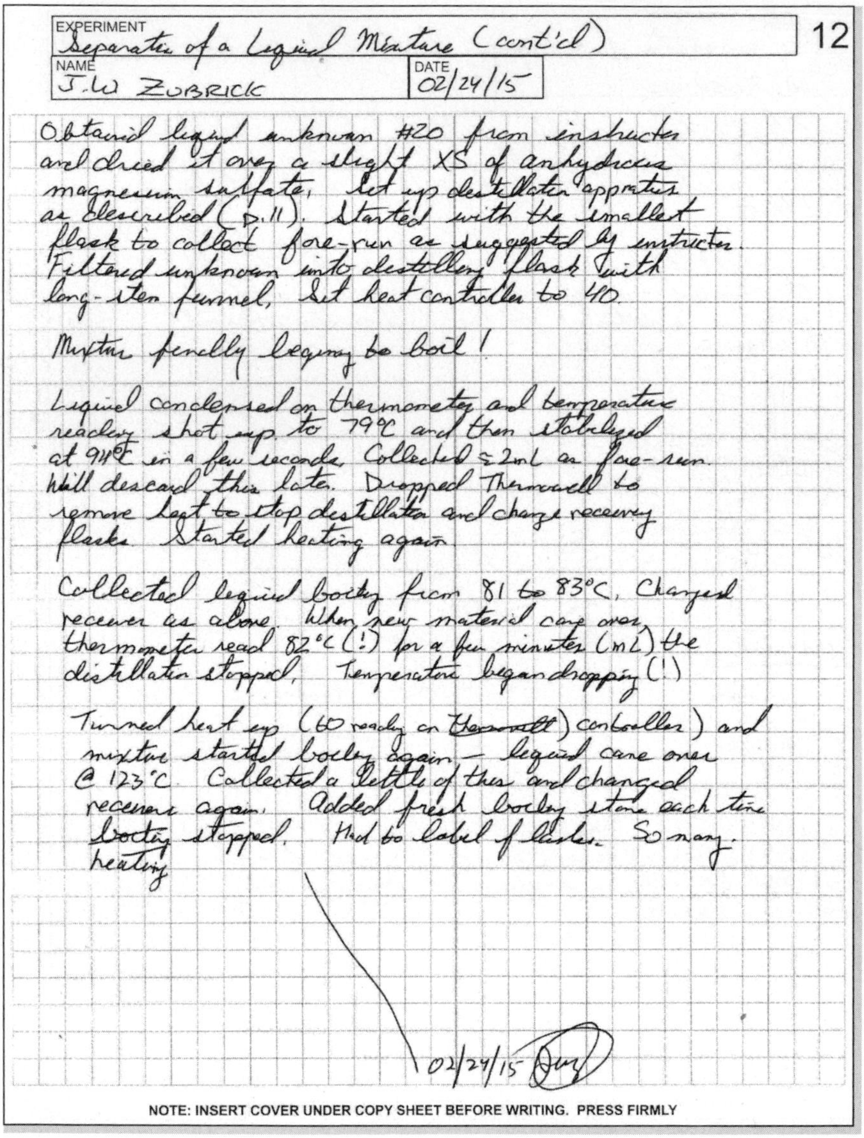

EXPERIMENT
Separation of a Liquid Mixture (cont'd)   12

NAME
J.W. ZUBRICK

DATE
02/24/15

Obtained liquid unknown #20 from instructor
and dried it over a slight XS of anhydrous
magnesium sulfate. Set up distillation apparatus
as described (p.11). Started with the smallest
flask to collect fore-run as suggested by instructor.
Filtered unknown into distilling flask with
long-stem funnel. Set heat controller to 40.

Mixture finally begins to boil!

Liquid condensed on thermometer and temperature
reading shot up to 79°C and then stabilized
at 91°C in a few seconds. Collected ≈ 2 mL as fore-run.
Will descard this later. Dropped Thermowell to
remove heat to stop distillation and change receiving
flasks. Started heating again.

Collected liquid body from 81 to 83°C. Changed
receiver as above. When new material came over,
thermometer read 82°C (!) for a few minutes (ml) the
distillation stopped. Temperature began dropping (!)

Turned heat up (60 reads on Thermowell) controller) and
mixture started boiling again — liquid came over
@ 123°C. Collected a bottle of this and changed
receivers again. Added fresh boiling stone each time
boiling stopped. Had to label flasks. So many.
heating

02/24/15 [signature]

FIGURE 2.2 Notebook entry for a technique experiment (2).

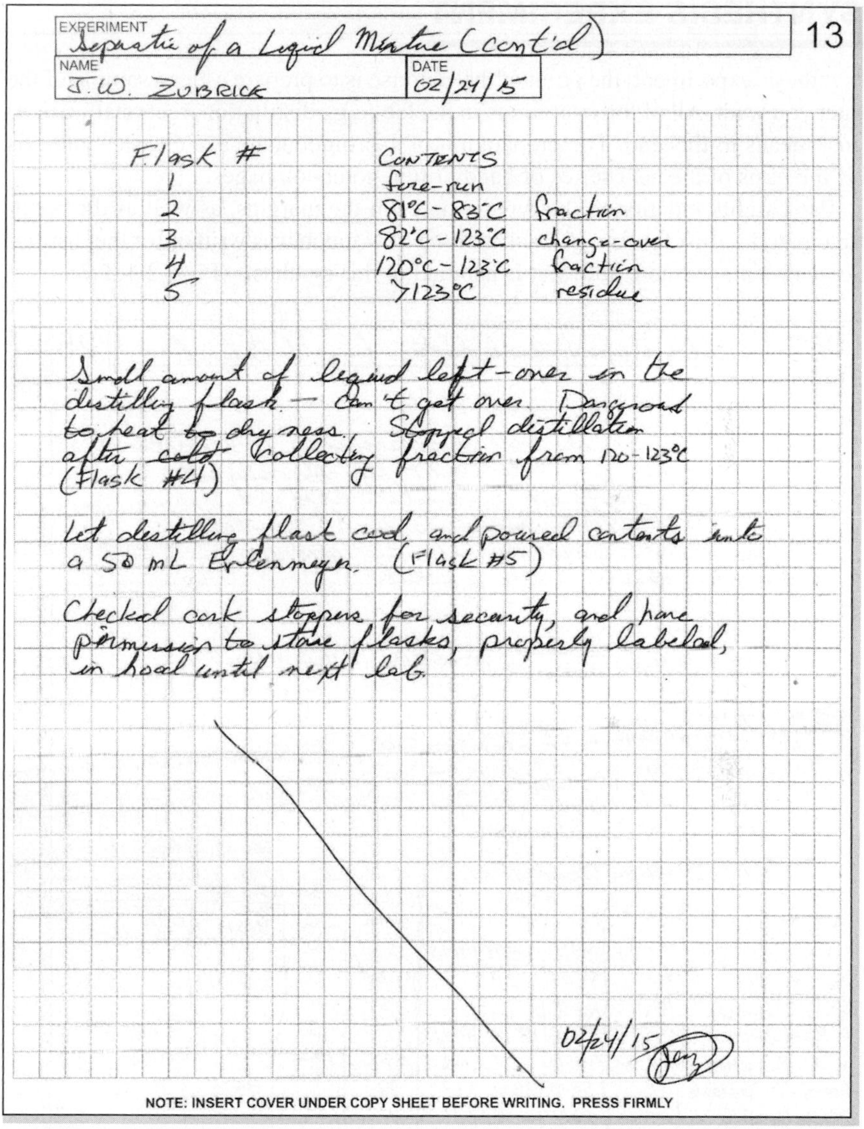

EXPERIMENT
Separation of a Liquid Mixture (cont'd)   13

NAME
J.W. ZUBRICK

DATE
02/24/5

Flask #	CONTENTS	
1	fore-run	
2	81°C - 83°C	fraction
3	82°C - 123°C	change-over
4	120°C - 123°C	fraction
5	>123°C	residue

Small amount of liquid left-over in the
distilling flask — can't get over. Dangerous
to heat to dryness. Stopped distillation
after cold collecting fraction from 120-123°C
(Flask #4)

Let distilling flask cool and poured contents into
a 50 mL Erlenmeyer. (Flask #5)

Checked cork stoppers for security, and have
permission to store flasks, properly labeled,
in hood until next lab.

02/24/15 Jay

NOTE: INSERT COVER UNDER COPY SHEET BEFORE WRITING. PRESS FIRMLY

FIGURE 2.3   Notebook entry for a technique experiment (3).

A SYNTHESIS EXPERIMENT

In a synthesis experiment, the point of the exercise is to prepare a clean sample of the product you want. All of the operations in the lab (e.g., distillation, recrystallization) are just means to this end. The preparation of 1-bromobutane is a classic synthesis and is the basis of the next series of handwritten notebook pages.

Pay careful attention to the typeset notes in the margins, as well as the handwritten material. Just for fun, go back and see how much was written for the distillation experiment, and note how that is handled in this synthesis (Figs. 2.4–2.7).

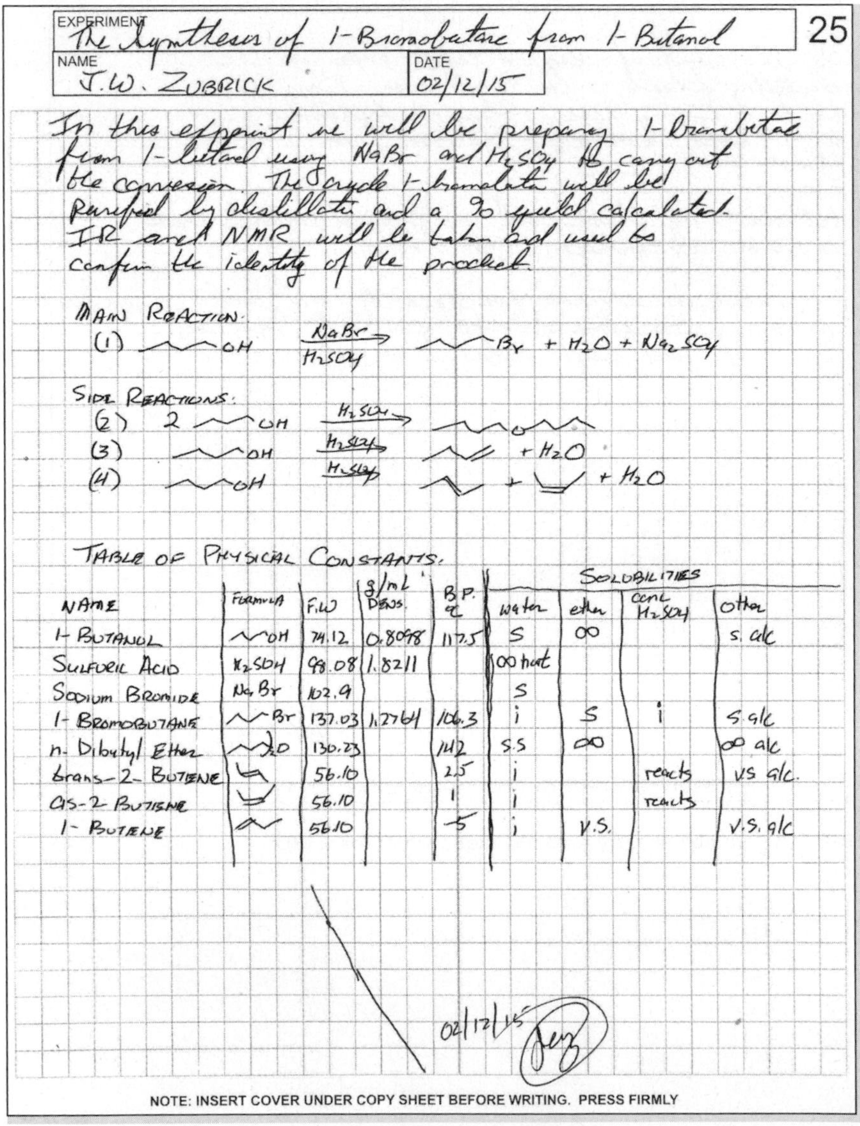

EXPERIMENT
The Synthesis of 1-Bromobutane from 1-Butanol   25

NAME   J.W. ZUBRICK   DATE 02/12/15

In this experiment we will be preparing 1-bromobutane from 1-butanol using NaBr and H_2SO_4 to carry out the conversion. The crude 1-bromobutane will be purified by distillation and a % yield calculated. IR and NMR will be taken and used to confirm the identity of the product.

MAIN REACTION:

(1) ... OH + NaBr / H_2SO_4 → ... Br + H_2O + Na_2SO_4

SIDE REACTIONS:

(2) 2 ... OH + H_2SO_4 → ... O ...

(3) ... OH + H_2SO_4 → ... + H_2O

(4) ... OH + H_2SO_4 → ... + ... + H_2O

TABLE OF PHYSICAL CONSTANTS:

NAME	FORMULA	F.W.	DENS. (g/mL)	B.P. °C	Water	ether	conc H_2SO_4	other
1-BUTANOL	...OH	74.12	0.8098	117.5	S	∞		s. alc
SULFURIC ACID	H_2SO_4	98.08	1.8211		∞ hot			
SODIUM BROMIDE	NaBr	102.9			S			
1-BROMOBUTANE	...Br	137.03	1.2764	106.3	i	S	i	s. alc
n-DIBUTYL ETHER	...O	130.23		142	s.s	∞		∞ alc
trans-2-BUTENE		56.10		2.5	i		reacts	vs alc.
cis-2-BUTENE		56.10		1	i		reacts	
1-BUTENE		56.10		-5	i	v.s.		v.s. alc

SOLUBILITIES

02/12/15

NOTE: INSERT COVER UNDER COPY SHEET BEFORE WRITING. PRESS FIRMLY

FIGURE 2.4   Notebook entry for a synthesis experiment (1).

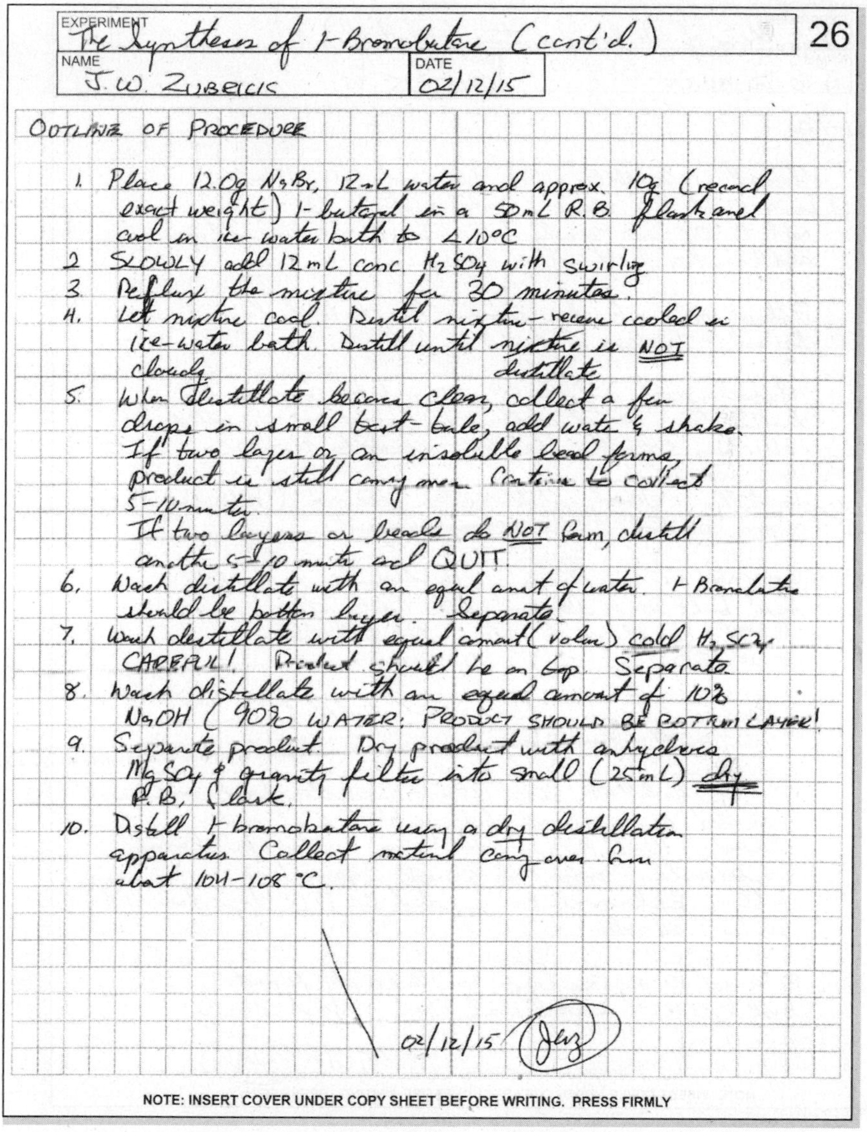

EXPERIMENT The Synthesis of 1-Bromobutane (cont'd.) 26

NAME J. W. Zubrick DATE 02/12/15

OUTLINE OF PROCEDURE

1. Place 12.0g NaBr, 12mL water and approx. 10g (record exact weight) 1-butanol in a 50mL R.B. flask and cool in ice-water bath to < 10°C
2. SLOWLY add 12 mL conc. H_2SO_4 with swirling
3. Reflux the mixture for 30 minutes.
4. Let mixture cool. Distill mixture – receiver cooled in ice-water bath. Distill until mixture is NOT cloudy distillate
5. When distillate becomes clear, collect a few drops in small test-tube, add water & shake. If two layers or an insoluble bead forms, product is still coming over. Continue to collect 5–10 minutes. If two layers or beads do NOT form, distill another 5–10 minutes and QUIT.
6. Wash distillate with an equal amount of water. 1-Bromobutane should be bottom layer. Separate.
7. Wash distillate with equal amount (volume) cold H_2SO_4 CAREFUL! Product should be on top. Separate.
8. Wash distillate with an equal amount of 10% NaOH (90% WATER: PRODUCT SHOULD BE BOTTOM LAYER!
9. Separate product. Dry product with anhydrous $MgSO_4$ & gravity filter into small (25mL) dry R.B. flask.
10. Distill 1-bromobutane using a dry distillation apparatus. Collect material coming over from about 104–108 °C.

02/12/15 JWZ

NOTE: INSERT COVER UNDER COPY SHEET BEFORE WRITING. PRESS FIRMLY

FIGURE 2.5 Notebook entry for a synthesis experiment (2).

Once again, if your own instructor wants anything different, do it. The art of notebook keeping has many schools—follow the perspective of your own school.

Notebook Notes

1. Use a descriptive title for your experiment. "n-Butyl Bromide." So what? Did you drink it? Set it on fire? What?! "The Synthesis of 1-Bromobutane from 1-Butanol"—now *that's* a title.
2. Do you see a section for unimportant side reactions? No. Then don't include any.

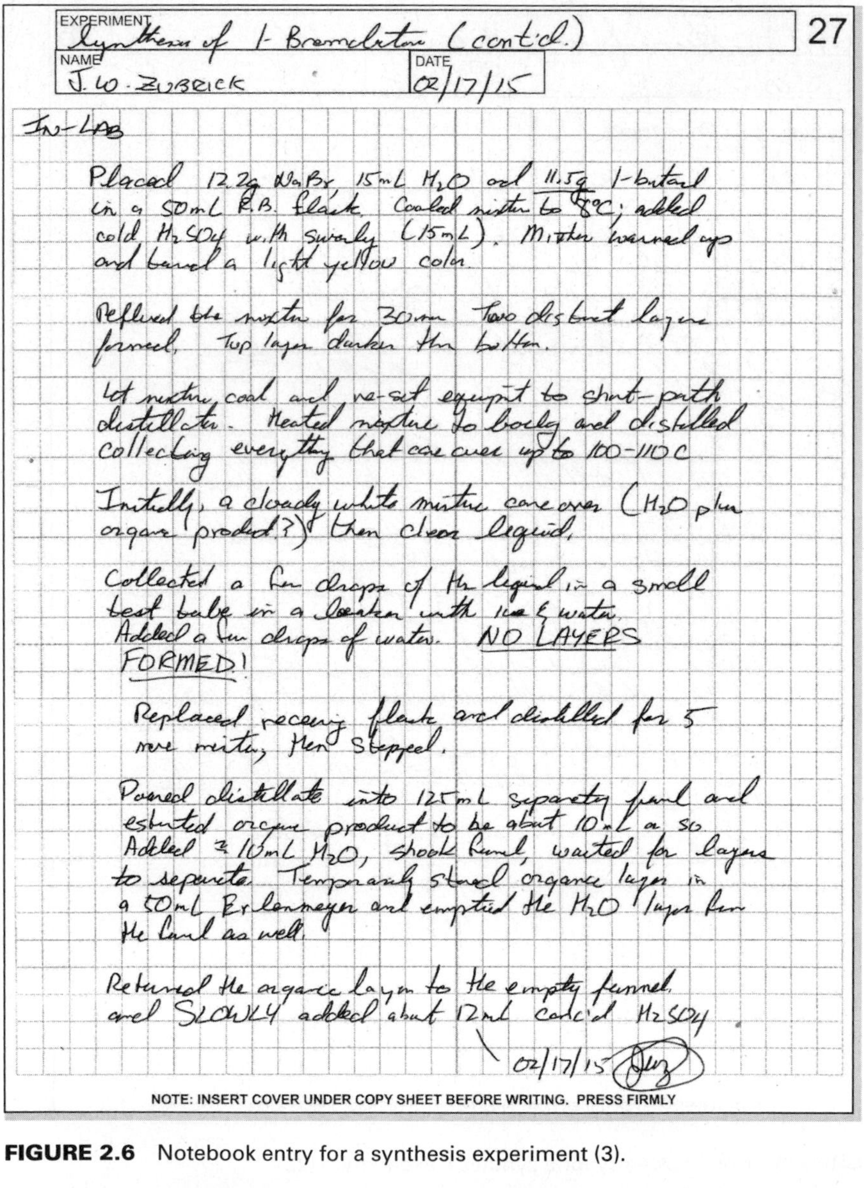

EXPERIMENT Synthesis of 1-Bromobutane (cont'd.) 27

NAME J. W. ZUBRICK   DATE 02/17/15

In-Lab

Placed 12.2g NaBr, 15mL H_2O and 11.5g 1-butanol in a 50mL R.B. flask. Cooled mixture to 0°C; added cold H_2SO_4 with swirly (15mL). Mixture warmed up and turned a light yellow color.

Refluxed the mixture for 30min. Two distinct layers formed. Top layer darker than bottom.

Let mixture cool and re-set equipment to short-path distillation. Heated mixture to boiling and distilled collecting everything that came over up to 100-110°C

Initially, a cloudy white mixture came over (H_2O plus organic product?) then clear liquid.

Collected a few drops of the liquid in a small test tube in a beaker with ice & water. Added a few drops of water. NO LAYERS FORMED!

Replaced receiving flask and distilled for 5 more minutes, then stopped.

Poured distillate into 125mL separatory funnel and estimated organic product to be about 10mL or so. Added ≈10mL H_2O, shook funnel, waited for layers to separate. Temporarily stored organic layer in a 50mL Erlenmeyer and emptied the H_2O layer from the funnel as well.

Returned the organic layer to the empty funnel and SLOWLY added about 12mL conc'd H_2SO_4

02/17/15

NOTE: INSERT COVER UNDER COPY SHEET BEFORE WRITING. PRESS FIRMLY

FIGURE 2.6 Notebook entry for a synthesis experiment (3).

3. In this experiment, we use a 10% aqueous sodium hydroxide solution as a wash (see Chapter 15, "Extraction and Washing") and anhydrous calcium chloride as a drying agent (see Chapter 10, "Drying Agents"). These are not listed in the Table of Physical Constants. They are neither reactants nor products. Every year, however, somebody lists the physical properties of *solid* sodium hydroxide, calcium chloride drying agent, and a bunch of other reagents that have nothing to do with the main synthetic reaction. I'm especially puzzled by a listing of solid sodium hydroxide in place of the 10% solution.

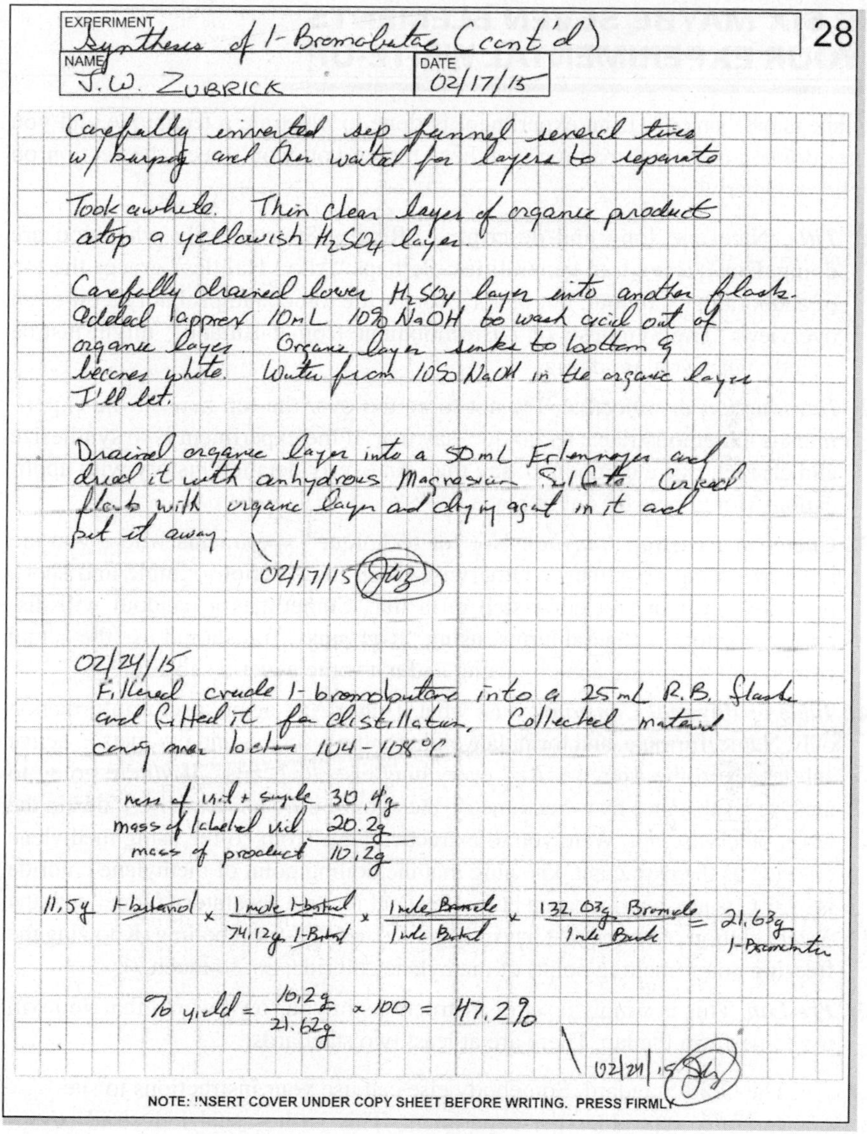

EXPERIMENT: Synthesis of 1-Bromobutane (cont'd) **28**

NAME: J.W. ZUBRICK   DATE: 02/17/15

Carefully inverted sep funnel several times w/ burping and then waited for layers to separate

Took awhile. Thin clear layer of organic product atop a yellowish H_2SO_4 layer

Carefully drained lower H_2SO_4 layer into another flask. Added approx 10mL 10% NaOH to wash acid out of organic layer. Organic layer sunk to bottom & becomes white. Water from 10% NaOH in the organic layer. I'll bet.

Drained organic layer into a 50mL Erlenmeyer and dried it with anhydrous Magnesium Sulfate. Corked flask with organic layer and drying agent in it and put it away

02/17/15 (JWZ)

02/24/15
Filtered crude 1-bromobutane into a 25mL R.B. flask and fitted it for distillation. Collected material coming over 104-108 °C

mass of vial + sample 30.4g
mass of labeled vial   20.2g
mass of product   10.2g

$$11.5 g\ 1\text{-butanol} \times \frac{1\ mole\ 1\text{-butanol}}{74.12g\ 1\text{-Butanol}} \times \frac{1\ mole\ Bromide}{1\ mole\ Butanol} \times \frac{132.03g\ Bromide}{1\ mole\ Butanol} = 21.63g\ 1\text{-Bromobutane}$$

$$\% \text{ yield} = \frac{10.2g}{21.62g} \times 100 = 47.2\%$$

02/24/15 (JWZ)

NOTE: INSERT COVER UNDER COPY SHEET BEFORE WRITING. PRESS FIRMLY

FIGURE 2.7   Notebook entry for a synthesis experiment (4).

4. I'm a firm believer in the use of units, factor-label method, dimensional analysis, whatever you call it. I know I've screwed up if my units are (g 1-butanol)^2/mole 1-butanol

5. Remember the huge write-up on the "Separation of a Liquid Mixture by Distillation," drawings of apparatus and all? Just a very few words (Fig. 2.7) are all you need to write for a distillation during this synthesis.

THE SIX MAYBE SEVEN ELEMENTS IN YOUR EXPERIMENTAL WRITE-UP

Why six maybe seven? If an experiment is done to illustrate a technique and you are not doing a chemical reaction (synthesis), then you won't need the section on Reactions. Otherwise:

1. *Title.* Nice and long and descriptive. Really. State exactly what you are doing. Don't leave it to generalities. Perhaps "Alkyl Halides" graces the top of a laboratory handout, but that's a very wide topic, and probably a bit of an overview. "The Synthesis of 1-Bromobutane from 1-Butanol" might just be exactly what you are doing.

2. *Introduction or objective.* Try not to be too over-the-top here. If the experiment is to demonstrate a technique, say that. If the experiment is to synthesize and characterize a compound, say that. An overly detailed history with applications would be a bit out of place.

3. *Chemical Reaction (maybe).* Not for technique experiments where you are not performing a synthesis. Otherwise, keep the reactions simple and short, preferably just one reaction step on a line. Sometimes a handout will discuss a reaction in general terms, using "R-groups." You should use the actual compound(s) you are actually using and not some generic substitute.

4. *Table of Physical Constants.* Just what it says. And again, useful information only. Name, formula, and formula weight for the reactants and products or active substances in the exercise. And everything else "depends." If you're going to analyze a substance by refractometry, then you'll need its refractive index in the table; otherwise not. Were you to extract caffeine from coffee using methylene chloride as the extractant, knowing that the boiling point of methylene chloride is 39.6°C is terribly important before you put it into a separatory funnel with the water solution of coffee that has only cooled to 50°C from boiling. Knowing the freezing point (melting point) of methylene chloride, not so much.

5. *Pre-Lab.* This is simply a set of instructions in your own words that you will use to perform the lab. There are at least two standards:

   The Gold Standard. Somebody else can use your instructions to successfully reproduce the experiment. This is the usual industrial or academic research standard where a project might be worked on by several people over some period of time.

   The Silver Standard. You, yourself reading only your own instructions—no other laboratory manuals or handouts—can successfully complete the experiment. This allows for considerable leeway in the number of steps, amount of detail, and all that.

6. *In-Lab.* This is what you write in your notebook recording what you've done and what you see happen as the experiment proceeds. There will be many entries where what you write will be very close to the pre-lab instructions you've written. Try to keep your notes current. Do a little; write a little. Take your notebook with you to the balance, to any instrumentation, to any place you'll be recording data. No loose pieces of paper. Ever.

7. *Post-Lab.* Calculations, musings, conclusions, and anything else pertinent to the experiment.

THE ACID TEST

After all of this, you're still not sure what to write in your notebook? Try these simple tests at home.

1. Before lab. "Can I carry out this experiment without any lab manual?"

2. After lab. (I mean *immediately* after; none of this "I'll write my observations down later" garbage.) Ask yourself: "If this were someone else's notebook, could I duplicate the results exactly?"

If you can truthfully answer "yes" to these two questions, you're doing very well indeed.

NOTEBOOK MORTAL SIN

"Bless me, Father, for I have sinned."

"I let someone else borrow my notebook, and he seems to have dropped off the face of the earth. He doesn't come to lab or lecture anymore, and friends can't seem to find him either. My notebook is gone!"

"Well, for your penance . . . I'd say losing your lab notebook is punishment enough."

This doesn't happen often, but it's happened often enough. Lending your lab notebook to anyone else means you really don't have the slightest idea of the purpose of a laboratory notebook. It is not a hand-in project graded for neatness and completeness at the end of a semester. It is the place for you to make notes to yourself about what you are going to do in lab that day and the place you make notes to yourself about what you did and observed in lab that day, too.

What is *another* person going to do with your notebook? Copy your notes made before lab so he—not having even read the lab—could be dangerous not knowing what he's doing? Copy your notes made during the lab so he has the same data and observations you have about an experiment he hasn't done, leeching off your work?

Keep your notebook, and keep it to yourself.

CALCULATION OF PERCENT YIELD (NOT YEILD!)

1. *Percent recovery.* When you don't (or can't) know how much you "should get," you're stuck with simply dividing how much you started with into how much you got, and multiplying by 100.

$$\% \text{ Recovery} = \frac{\text{mass of material recovered in grams}}{\text{mass of materials started with in grams}} \times 100$$

If you're isolating a natural product—say, caffeine from coffee—there isn't any way you can possibly know exactly how much caffeine is actually in the sample of coffee you are working on. Yes, you can look up averages and typical amounts and make real good guesses, but they don't necessarily apply to your sample of coffee. Incidentally, the percent recoveries for natural products are exceedingly small; don't panic.

If you're separating and isolating an "unknown mixture," you can't know the masses of the components—that would be telling.

2. *Percent yield.* When you know, or can calculate, how much you "should get," then percent yield is your guy. The actual yield is the easy part—just the weight in grams of your product. The sticking point is usually calculating how much product you'd get if both the reaction and your technique were perfect—the theoretically perfect attainable yield. Since this is just a paper calculation, some call it a calculated yield rather than the more common term, theoretical yield.

$$\% \text{ Yield} = \frac{\text{actual yield of product in grams}}{\text{theoretical yield in grams}} \times 100$$

To calculate the theoretical yield, you have to have a balanced equation, and because chemical reactions happen on a molecular basis, yep, you have to calculate the number of moles of some of the substances you have, determine the limiting reagent (that again!), and use the molecular weights of the compounds involved. Hey, that's why you looked them up and wrote them down in the Table of Physical Constants in your notebook, eh?

Let's look at the reaction and calculations for that synthesis of 1-bromobutane. First and foremost—and this takes a bit of experience (ouch)—the only reagent that is important here in making the product is the 1-butanol. Yes, you need sulfuric acid. Yes, you need sodium bromide. But look at the conversion you're trying to do, and look at the product you're trying to make: 1-butanol to 1-bromobutane. Unless the reaction conditions have been set up very oddly indeed, the limiting reagent will be the important organic compound, here 1-butanol. The sulfuric acid and sodium bromide will be in molar excess (often abbreviated XS). So all of the calculations are based on the quantity of 1-butanol. Of course, the most cautious thing to do is to calculate the number of moles of everything and, taking into account any stoichiometric factors, base your calculations on that.

Because the quantity of 1-butanol is given as a volume, you have to use the density to convert it to a mass. Then, looking up the molecular weight of 1-butanol in the Table of Physical Constants from your notebook, use that to convert the mass of 1-butanol to the number of moles of 1-butanol. Using the stoichiometric factor—here every 1 mole of 1-butanol gives 1 mole of 1-bromobutane—calculate the number of moles of 1-bromobutane. Finally, using the molecular weight for 1-bromobutane (also in your Table), calculate the theoretical yield, in grams, of 1-bromobutane. It's probably best to use the factor label (or "units") method to do these calculations, and see that the units lead to the correct result as well as the numbers. This is all laid out in Figure 2.5. Note that there isn't even an attempt to calculate the moles of sodium bromide or sulfuric acid, though.

In that example, we begin with 17.0 mL of 1-butanol, and wind up with 16.2 g of the product, 1-bromobutane. To get the percent yield, you have to:

1. *Convert the volume of product into the mass.* You've obtained the 1-butanol as a liquid, measured as a volume. Yes, you can weigh liquids, but we didn't here. Volume-to-mass (and mass-to-volume) conversions use the density. You want to go to mass (g), and you want to get rid of the volume, so

$$17.0 \text{ mL 1-butanol} \times \frac{0.8098 \text{ g 1-butanol}}{1 \text{ mL 1-butanol}} = 13.77 \text{ g 1-butanol}$$

2. *Calculate the moles of 1-butanol.* Here you use the molecular weight. It's given in g/mol, but you have to flip that relationship. If you don't, your units will wind up being g^2/mol instead of mol.

$$13.77 \text{ g 1-butanol} \times \frac{1 \text{ mol 1-butanol}}{74.12 \text{ g 1-butanol}} = 0.1857 \text{ mol 1-butanol}$$

3. *Use the stoichiometric factor to get moles of 1-bromobutane.* Here, 1 mole of 1-butanol is converted to 1 mole of 1-bromobutane. Even though this is pretty simple, keep the units anyway. There will be times when the reaction is not 1:1, and if you don't get the fraction set up correctly—that's all, folks.

$$0.1857 \text{ mol 1-butanol} \times \frac{1 \text{ mol 1-bromobutane}}{1 \text{ mol 1-butanol}} = 0.1857 \text{ mol 1-bromobutane}$$

4. *Calculate the mass of 1-bromobutane.* Now you use the molecular weight of 1-bromobutane, as stated. If you've done everything right, the units work out to be grams of 1-bromobutane.

$$0.1857 \text{ mol 1-bromobutane} \times \frac{137.08 \text{ g 1-bromobutane}}{1 \text{ mol 1-bromobutane}}$$
$$= 25.44 \text{ g 1-bromobutane}$$

5. *Calculate the percent yield.* This is the amount of product you actually obtained, here 16.2 g, divided by the calculated (theoretical) yield of 25.44 g, multiplied by 100.

$$\frac{16.2 \text{ g 1-bromobutane obtained}}{25.44 \text{ g 1-bromobutane calculated}} \times 100 = 63.7\%$$

You usually report this calculated percent yield in whole numbers. Here, 63.7% rounds up to 64%.

6. *Watch out for significant figures.* Chances are you have a number like 63.679245 in your calculator window. You weighed your product to one part in ten (± 0.1), and calculated to one part in one hundred (± 0.01). If the product weight can vary by ± 0.1 g, there's no need for more figures than that.

ESTIMATION IS YOUR FRIEND

You've been asked to weigh 0.07 mole of a reagent called toluene into a 100-mL flask. You punch in the molecular weight and the 0.07, press the key, and get 25.2976496 g. Is this reasonable? Well, the molecular weight of toluene is 92.141 g/mol, and 0.07 is less than 0.1 and 0.1 times about 92 is about 9.2, so the result should have been less than 9.2, not anywhere near 25.

EXERCISES

1. Should what you've written in your notebook in the lab as you are doing the experiment look a lot like what you have written in your notebook as information to help you perform the lab? Why or why not?

2. A schedule of experiments for a laboratory indicates that the 1-bromobutane preparation is paired with the preparation of a tertiary halide, 2-methyl-2-chloropentane, under the single title "Preparation of Alkyl Halides." What do you say to someone who does not have a separate title, introduction, prelab section including a table of physical constants, and such for the preparation of this compound—even though he argues that the experiment schedule doesn't say to read a separate experiment?

3. Drag out your general chemistry book and solve any four stoichiometry problems dealing with a limiting reagent.

4. Unless otherwise specified, solutions are called "weight-weight," even though there may be more than two components, and masses will be used. Is a 10% sodium hydroxide solution made by dissolving 10 grams of the solid in 100 mL of water? Why or why not? What's up with insisting you should dissolve the 10 g of solid hydroxide in only 90 g of water?

MINING YOUR OWN DATA

You should look up information concerning any organic chemical you'll be working with so that you know what to expect in terms of molecular weight, density, solubility, crystalline form, melting or boiling point, color, and so on.

Up until fairly recently, if you wanted to get absolutely accurate data about the compounds you'd be working with, you'd go to a library and look up the information in chemistry handbooks. I suppose you could still do that, but it's more than likely you'll use your cell phone or other computer to get that data off of the Internet. Please note that I did not include the descriptors "absolutely accurate" in front of the word "data" this time. OK,OK ... My fondness for handbooks has gotten the better of me. Just be careful, by perhaps checking out more than one source, until you're confident in the quality of the data.

And that's if you can get the data at all. In the previous edition, I listed chemfinder.cambridgesoft.com as a place to get physical constants on compounds. Well, they've been bought by the Perkin-Elmer instrument company, and this time, my search on the compound 1-bromobutane—even after registering—threw me to ChemBioFinder.com, and gave me a picture, the name (and synonyms), the molecular weight (super-purists say "molar mahss"), but nothing anybody could use in the organic chemistry laboratory.

Because these kinds of "here-they-were-now-they're-gone" or "it-was-free-now-it's-pay" activities among the Internet, it's just best to stick to a search engine to get you hits for different websites where you can get the data you need.

GOOGLE AND THE WIKI

If I bring up the Google search page, and enter 1-bromobutane, the first entry on my list is to a Wikipedia entry, followed by PubChem, Sigma-Aldrich, and ChemSpider. Each has its own data and more importantly **data emphasis** than the others. If you're trying to fill out a Table of Physical Constants in your notebook, the Wikipedia entry will probably be the best. Understand, though, that whichever you choose, they will present you with way more information than you'll ever need.

The chemical compound Wikipedia pages are divided into three parts: at the left, a general index into the entire Wikipedia, text (called an article) with information about the compound, and finally, at the right, an extensive table of physical properties.

First there are some structures. Usually a bond-line formula representation along with a 3D-ish drawing of the molecule showing the orientation of the atoms. Look up triphenylmethanol. The 3D-ish drawing shows not all the rings are in the same plane as shown in the bond-line representation above it.

Next some names. The IUPAC (International Union of Pure and Applied Chemistry) name, often called the systematic name, followed by any other names and synonyms.

The section on identifiers will probably be a bit less useful. The CAS Registry Number is a unique identifier assigned by the Chemical Abstracts Service for that particular compound. If you had nothing but the CAS number on a scrap of paper, you could easily look up that compound alone in pretty much any database. And if what you're looking at has a different CAS number, be assured that they are NOT the same compound.

Finally physical properties. These can be extensive, as for the aforementioned 1-bromobutane, really, really extensive as for something like benzoic acid that has a lot of uses and a long history, or, as we say, not so much for something like triphenylmethanol, whose sole claim to fame appears to be having been prepared in hundreds of undergraduate organic chemistry laboratories to illustrate something called the Grignard reaction.

Some of the physical properties you might need:

Molecular formula: The kinds and numbers of the atoms that make up the compound in a simple, whole number ratio, often called the empirical formula.

Molar mass: The weight of 1 mole in grams: g/mol. You'll need this to calculate product yield in your syntheses.

Appearance: Makes its appearance for solids and describes the solid's crystalline form and color. Although I haven't yet seen any entry with abbreviations, crystalline forms are often: pl, plates; nd, needles; lf, leaves; mcl, monoclinic; rb, rhombus, and so on. Colors are usually two-letter abbreviations such as ye for yellow; pa for pale.

Liquids, if colored, can take on the names and abbreviations of colors. Liquids without color are always called colorless, and you never use the word clear. The antonym for clear is cloudy, and if you think about it, you could have a colorless liquid that is cloudy (often also called "milky"), or a colored liquid that is clear. Is that clear?

Density: Given in $g \cdot mL^{-1}$ and often as g/mL. That the density can vary with temperature means you have to find out at what temperature the density measurement was taken. Wikipedia insists all its temperature-sensitive data was taken at 25°C. Often a density taken at a temperature different from the posted standard will have the differing temperature as a superscript: for 1-bromobutane at 20°C not 25°C, 1.2758^{20} And sometimes the volume is given as cm^3 or cc for cubic centimeter. If you don't weigh a liquid reactant to get the mass directly, but use a volume of a liquid in your synthesis, you'll need the density data to convert the volume to grams to calculate your yield.

Melting point: Clear enough. Wikipedia entries sometimes state a range but mostly state a single number (in C and F and K, Oh My!). Usually the single number is the top number of a narrow (less than 2°C) melting point range. You should report the entire range in your notebook.

Boiling point: The boiling point at 100 kPa unless otherwise noted. Weird. The standard atmosphere is 101.325 kPa, and using exactly 100 kPa (1 bar) as a standard means there had to be some correction for pretty much all the liquids in the Wikipedia, as previous work was usually corrected, if it even need be, to 1 atmosphere, or 760 Torr.

Aside from the careful use of capillary tubes, you take the boiling point of any of your liquids in your lab any time you distill them. If you distill and collect a liquid from 81 to 83°C, put that in your notebook because that's the boiling point (range) of that liquid. As for solids, you can collect liquids over a small boiling range and report the upper number as the boiling point, making sure you have the actual range in your notebook.

Boiling points taken at pressures differing from that posted have a superscript much like densities, only it's the pressure in Torr the distillation was done at. So for 1-bromobutane, 18.8^{30} means it boils at 18.8°C at 30 Torr.

Refractive index (n_D): The index of refraction (see Chapter 29, "Refractometry") obtained using the yellow light from a sodium vapor lamp (the D line). That's the "D" in n_D.

Solubility in water … and other solvents, too: Solubility data presentation is all over the map running from way too much to none at all. Benzoic acid has entries for solubility in water, methanol, and ethanol at various temperatures, solubility in acetone, solubility in 1,4-dioxane, and even solubility in olive oil, of all things.

The values are given in g/100 g indicating the mass of the compound that dissolves in 100 g of the solvent. Not 100 mL, but 100 g. And there's even just a plain listing without any numeric data of the other things benzoic acid is soluble in. There is no solubility data for triphenylmethanol in the Wikipedia, so a trip to Google is in order.

Searching for triphenylmethanol solubility pulls up the same Wikipedia link that actually has NO solubility data (imagine that), but the second entry to something called ChemicalBook is more fruitful: Solubility in water: INSOLUBLE. Nothing but the one word, but it would be useful if you want to wash an ether solution of triphenylmethanol with water or a water solution of something, and know that the compound itself won't go into the wash water.

Although there seems to be enough room to use entire words in the solubility descriptions, there are fairly standard abbreviations that have been used for a long time. Some of them are as follows:

al	ethyl alcohol	eth	ethyl ether
bz	benzene	chl	chloroform
peth	petroleum ether	w	water
HOAc	acetic acid	MeOH	methyl alcohol
lig	ligroin	ctc	carbon tetrachloride
tol	toluene	ace	acetone
s	soluble	i	insoluble
sl, δ	slightly soluble	msc, ∞	miscible, mixes in all proportions
h	solvent must be hot	v	very

Some solvents have such a long tradition of use that they are our old friends and we use very informal names for them:

alcohol	ethyl alcohol; ethanol
pet. ether	petroleum ether; not a true ether, but a low-boiling (30–60°C) hydrocarbon fraction like gasoline
ether	diethyl ether; ethoxyethane.
ligroin	another hydrocarbon mixture with a higher boiling range (60–90°C) than pet. ether.

Thermochemistry: Sometimes there is thermodynamic data, and sometimes there is not. And unless you're doing physical organic chemistry, it won't be very useful.

Hazards: I was going to say that every compound had at least a Material Safety Data Sheet (MSDS) detailing the precautions you should take, but for the compound terphenyl, the active ingredient in the old Lionel Trains smoke pellets, there is no listing. Otherwise, you get more safety data than you may ever need.

Spectra: Only occasionally are there links to the IR and NMR spectra. Better you should Google your compound with the type of spectrum you're looking for, example, 1-Bromobutane IR, and when the results come up, click on the word "Images" at the top of the page. Gives you actual spectra.

Those are some of the more common and more useful physical data available for chemical compounds from Wikipedia. Remember to be flexible in your thinking and searching, and to examine the data with a critical eye.

THE TERPHENYL ANOMALY

While this sounds like an episode of *The Big Bang Theory*, it illustrates just how careful, and flexible, and did I say careful you have to be searching for physical data on the Internet. Terphenyl, at its base, has three aromatic rings hooked together, but they can be hooked together in three different ways.

If you Google terphenyl, the Wikipedia entry that comes up has pictures of the three different isomers, but data for only one of them, *p*- or *para*- Terphenyl with the three rings in a straight line. And if you specifically Google *m*- or *meta*-Terphenyl or *o*- or *ortho*-Terphenyl, the Wikipedia entry that comes up is the same one for p-Terphenyl, and if you want data for the others, you are out of WikiLuck.

You have to have the presence of mind to realize first, that this Wikipedia entry has no data for the other isomers, so don't start immediately copying data from the page, and second, that you'll have to choose a different source from your search for your data.

If you have any questions about this Internet stuff or that chemistry stuff, you can e-mail me at

j.zubrick@hvcc.edu

EXERCISES

1. Look up the physical properties of several of the compounds you will be making in your lab from several different places on the Internet. Do you believe them all to be correct?

2. Look up the Freidel-Crafts acylation reaction of toluene and acetyl chloride. (This is generally Orgo II; stretch yourself!) Look up the product(s) in a handbook. Comment upon the fact that the product is not named as a substituted toluene, even though the acetyl group did substitute for a hydrogen on the toluene ring.

JOINTWARE

■ *Grease the joints to stop sticking.*

Using **standard taper jointware**, you can connect glassware without rubber stoppers, corks, or tubing. Pieces are joined by ground glass connections built into the apparatus (Fig. 4.1). They are manufactured in standard sizes, and you'll probably use ⊺19/22.

The symbol ⊺ means **standard taper**. The first number is the size of the joint at the widest point in millimeters. The second number is the length of the joint in millimeters. This is simple enough. Unfortunately, life is not all that simple, except for the mind that thought up this next devious little trick.

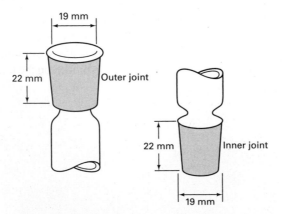

FIGURE 4.1   Standard taper joints (⊺19/22).

STOPPERS WITH ONLY ONE NUMBER

Sounds crazy, no? But with very little imagination, and even less thought, grave problems can arise from confusing the two. O.K., O.K. These days most of the single-numbered stoppers are plastic ones that look wildly different from the glass ones, but still, be careful about mixing them up. Look at Figure 4.2, which shows both glass and plastic single-number stoppers trying and failing to fit into standard taper joints. Intermixing a ⊤ 19/22 and a ⊤ 19 stopper leads to **leaking joints** through which your **graded** product can escape. Also, the ⊤ 19/22 stopper is much more expensive than the ⊤ 19 stopper, and you may *have to pay money* to get the correct one when you check out at the end of the course. Please note the emphasis in those last two sentences. I appeal to your better nature and common sense. So, take some time to check these things out.

As you can see from Figure 4.2, that single number is the width of the stopper at its top. There is no mention of the length, and you can see that it is too short. The ⊤ 19 stopper *does not* fit the ⊤ 19/22 joint. Only the ⊤ 19/22 stopper can fit the ⊤ 19/22 joint. Single-number stoppers are commonly used with volumetric flasks. Again, they will leak or stick if you put them in a double-number joint.

With these delightful words of warning, we continue the saga of coping with ground-glass jointware. Figure 4.3 shows some of the more familiar pieces of jointware that you may encounter in your travels. They may not be so familiar to you now, but give it time. After a semester or so, you'll be good friends, go to reactions together, maybe take in a good synthesis. Real fun stuff!

These pieces of jointware are the more common pieces that I've seen used in the laboratory. You may or may not have *all* the pieces shown in Figure 4.4. Nor will they necessarily be called by exactly the names given here. The point is, *find out* what each piece is, and *make sure* that it is in good condition *before* you sign your life away for it.

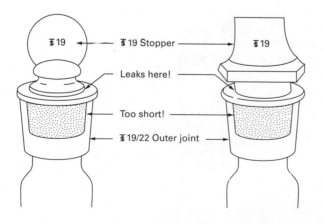

FIGURE 4.2 A ⊤ 19 nonstandard in ⊤ 19/22 standard taper joints.

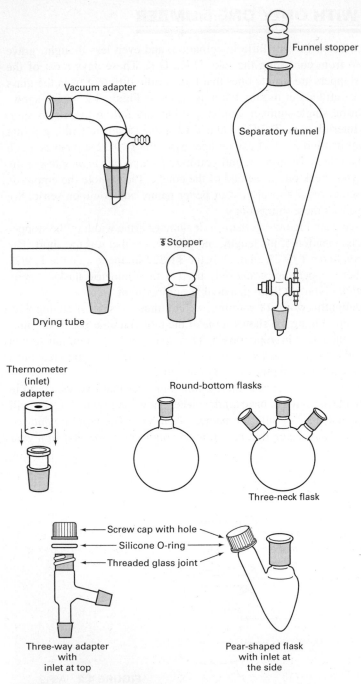

Vacuum adapter

Funnel stopper

Separatory funnel

Drying tube

Stopper

Thermometer
(inlet)
adapter

Round-bottom flasks

Three-neck flask

Screw cap with hole

Silicone O-ring

Threaded glass joint

Three-way adapter
with
inlet at top

Pear-shaped flask
with inlet at
the side

FIGURE 4.3 Some jointware.

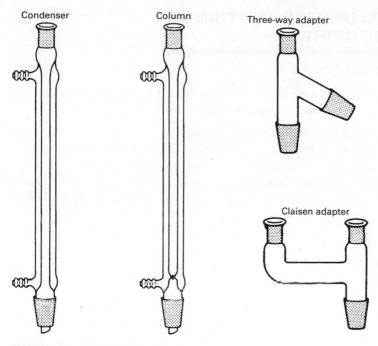

Condenser   Column   Three-way adapter

Claisen adapter

FIGURE 4.4   More jointware.

ANOTHER EPISODE OF LOVE OF LABORATORY

"And that's $95.48 you owe us for the separatory funnel."

"But it was broken when I got it!"

"Should've reported it then."

"The guy at the next bench said it was only a two-dollar powder funnel and not to worry and the line at the stockroom was long anyway, and . . . and . . . anyway, the stem was only cracked a little . . . and it worked OK all year long . . . Nobody said anything . . ."

"Sorry."

Tales like this are commonplace, and ignorance is no excuse. Don't rely on expert testimony from the person at the next bench. He may be more confused than you are. And equipment that is "slightly cracked" is much like a person who is "slightly dead." There is no in-between. If you are told that you *must* work with damaged equipment because there is no replacement available, you would do well to get it in writing.

HALL OF BLUNDERS AND THINGS NOT QUITE RIGHT

Round-Bottom Flasks

Round-bottom (RB) jointware flasks are so round and innocent looking that you would never suspect they can turn on you in an instant.

1. *Star cracks.* A little-talked-about phenomenon that turns an ordinary RB flask into a potentially explosive monster. Stress, whether prolonged heating in one spot or indiscriminate trouncing on hard surfaces, can cause a flask to develop a **star crack** (Fig. 4.5) on its backside. Sometimes a crack may be hard to see, but if it is overlooked, the cracked flask may split at the next lab.

Star crack

FIGURE 4.5   Round-bottom flask with star crack.

2. *Heating a flask.* Since they are cold-blooded creatures, flasks show more of their unusual behavior when they are being heated. The behavior is usually unpleasant if certain precautions are not taken. In addition to star cracks, various states of disrepair can occur, leaving you with a bench top to clean. Both humane and cruel heat treatment of flasks will be covered in Chapter 17, "Sources of Heat," which is on the SPCG (Society for the Prevention of Cruelty to Glassware) recommended reading list.

Columns and Condensers

A word about **distilling columns** and **condensers**:

Different!

Use the **condenser** as is for **distillation** and **reflux** (see Chapter 19, "Distillation," and Chapter 22, "Reflux and Addition"). You can use the column *with or without column packing* (bits of metal or glass or ceramic or stainless steel sponge—whatever!). That's why the column is wider and has *projections* at the ends (Fig. 4.6). These projections help hold up the column packing if you use any packing at all (see Figure 19.10).

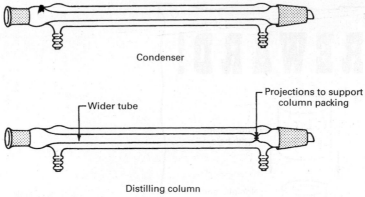

Condenser

Projections to support
column packing

Wider tube

Distilling column

FIGURE 4.6 Distilling column versus condenser.

If you jam column packing into the skinny condenser, the packing may never come out again! Using a condenser for a packed column is bad form and can lower your esteem or grade, whichever comes first.

> *You might use the column as a condenser.*
> *Never use the condenser as a packed column!*

The Adapter with Lots of Names

Figure 4.7 shows the one place where joint and nonjoint apparatus meet. There are two parts: a rubber cap with a hole in it and a glass body. Think of the rubber cap as a rubber stopper through which you can insert thermometers, inlet adapters, drying tubes, and so on. (There's a version that combines this guy with the three-way adapter. See Chapter 5, "Microscale Jointware.")

> *CAUTION! Do not force. You might snap the part you're trying to insert.*
> *Handle both pieces through a cloth; lubricate the part (water) and then*
> *insert carefully. Keep fingers on each hand no more than 2 in. apart.*

The rubber cap fits over the **nonjoint** end of the glass body. The other end is a **ground-glass joint** and *fits only other glass joints.* The rubber cap should neither crumble in your hands nor need a 10-ton press to bend it. If the cap is shot, get a new one. Let's have none of these corks, rubber stoppers, chewing gum, or any other type of plain vanilla adapter you may have hiding in the drawer.

And remember: Not only thermometers, but **anything** that resembles a glass tube can fit in here! This includes unlikely items such as **drying tubes** (they have an outlet tube) and even a **funnel stem** (you may have to couple the stem to a smaller glass tube if the stem is too fat).

The imaginative arrangements shown in Figure 4.8 are acceptable.

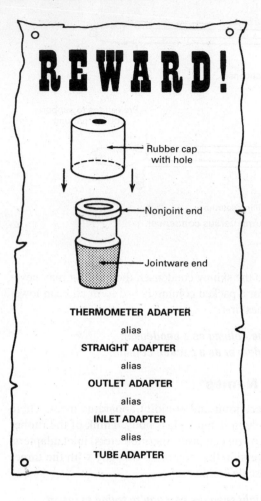

REWARD!

Rubber cap
with hole

Nonjoint end

Jointware end

THERMOMETER ADAPTER

alias

STRAIGHT ADAPTER

alias

OUTLET ADAPTER

alias

INLET ADAPTER

alias

TUBE ADAPTER

FIGURE 4.7 Thermometer adapter.

Forgetting the Glass (Fig. 4.9)

The Corning people went to a lot of trouble to turn out a piece of glass that fits perfectly in *both* a glass joint *and* a rubber adapter, *so use it!*

Inserting Adapter Upside Down

This one (Fig. 4.10) is really ingenious. If you're tempted in this direction, go sit in the corner and repeat over and over:

"Only glass joints fit glass joints."

SOCIALLY ACCEPTABLE THINGS TO DO WITH
THE ADAPTER WITH LOTS OF NAMES

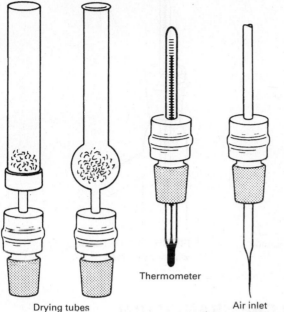

Drying tubes

Thermometer

Air inlet
(vacuum distillation)

FIGURE 4.8   Unusual,
yet proper, uses of the
adapter with lots of
names.

FIGURE 4.9   The glassless glass adapter.

Inserting Adapter Upside Down *sans* Glass

I don't know whether to relate this problem (Fig. 4.11) to glass-forgetting or upside-downness, since both are involved. Help me out. If I don't see you trying to use an adapter upside down without the glass, I won't have to make such a decision. So, don't do it.

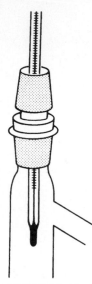

FIGURE 4.10 The adapter stands on its head.

FIGURE 4.11 The adapter on its head without the head.

THE O-RING AND CAP BRANCH OUT

Usually, you can find the O-ring, cap, and glass thread hanging out on your favorite pieces of microscale jointware (see Chapter 5, "Microscale Jointware"), but it can show up in some larger setups as well. Figure 4.3 shows a three-way adapter with this kind of inlet at the top, as well as a pear-shaped flask with an inlet at the side. You still have to be careful that you don't drop the rubber ring on the floor where you might not be able to see it very well, that you don't turn the cap so tightly you break the threads, and so on.

GREASING THE JOINTS

In all my time as an instructor, I've never had my students go overboard on greasing the joints, and they never got them stuck. Just lucky, I guess. Some instructors, however, use grease with a passion and raise the roof over it. The entire concept of greasing joints is not as slippery as it may seem.

To Grease or Not to Grease

Generally, you'll grease joints on two occasions: one, when doing vacuum work to make a tight seal that can be undone; the other, when doing reactions with a strong base that can etch the joints. Normally, you don't have to protect the joints during acid or neutral reactions.

Preparation of the Joints

Chances are you've inherited a set of jointware coated with 47 semesters of grease. First, wipe off any grease with a towel. Then soak a rag in any hydrocarbon solvent (hexane, ligroin, petroleum ether—and *no flames;* these burn like gasoline) and wipe the joint again. Wash off any remaining grease with a strong soap solution. You may have to repeat the hydrocarbon–soap treatments to get a clean, grease-free joint.

Some suggest using methylene chloride (CH_2Cl_2) to help remove silicone grease. Be very careful with this chlorinated hydrocarbon; dispose of it properly.

Into the Grease Pit (Fig. 4.12)

First, use only enough grease to do the job! Stipple grease all around the top of the inner joint. Push the joints together with a twisting motion. The joint should turn clear from one-third to one-half of the way down the joint. *At no time should the entire joint clear!* This means you have *too much grease* and you must start back at "Preparation of the Joints."

Don't interrupt the clear band around the joint. This is called **uneven greasing** and will cause you headaches later on.

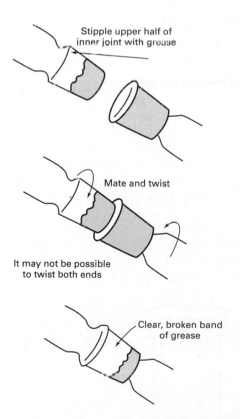

Stipple upper half of inner joint with grease

Mate and twist

It may not be possible to twist both ends

Clear, broken band of grease

FIGURE 4.12   Greasing ground-glass joints.

STORING STUFF AND STICKING STOPPERS

At the end of a grueling lab session, you're naturally anxious to leave. The reaction mixture is sitting in the joint flask, all through reacting for the day, waiting in anticipation of the next lab. You put the correct glass stopper in the flask, clean up, and leave.

The next time, the stopper is stuck!

Stuck but good! And you can probably kiss your flask, stopper, product, and grade goodbye!

Frozen!

Some material has gotten into the glass joint seal, dried out, and cemented the flask shut. There are few good cures but several excellent preventive medicines.

Corks!

Yes, corks; old-fashioned, non-stick-in-the-joint corks. If the material you have to store *does not attack cork,* this is the cheapest, cleanest method of closing off a flask.

A well-greased glass stopper *can* be used for materials that attack cork, but *only* if the stopper has a good coating of stopcock grease. Unfortunately, this grease can get into your product.

Do not use rubber stoppers!

Organic liquids can make rubber stoppers swell up like beach balls. The rubber dissolves and ruins your product, and the stopper won't come out either. Ever. The point is:

Dismantle all ground-glass joints before you leave!

CORKING A VESSEL

If winemakers corked their bottles the way some people cork their flasks, there'd be few oenophiles, and we'd probably judge good years for salad dressings rather than wines. You don't just take a new cork and stick it down into the neck of the flask, vial, or what have you. You must press the cork first. Then, as it expands, it makes a very good seal and doesn't pop off.

Before **pressing** or **rolling**, a brand-new cork should fit only about one-quarter of the way into the neck of the flask or vial. Then you roll the lower half of the cork on your *clean* bench top to soften and press the small end. *Now* stopper your container. The cork will slowly expand a bit and make a very tight seal (Fig. 4.13).

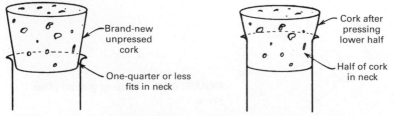

FIGURE 4.13 Corking a vessel.

MICROSCALE JOINTWARE

■ *Don't overtighten the caps.*

MICROSCALE: A FEW WORDS

Microscale is small. How small is it? Well, there's no absolute cutoff, but some suggest 10 g is macroscale, 1 g is semimicroscale, and 0.1 g is truly microscale. Now, 0.1 g of a **solid** is really something. A potassium bromide (KBr) infrared pellet uses 0.01 g (10 mg), and a nuclear magnetic resonance (NMR), which takes five times as much (50 mg), leaves you 40 mg to play with (find melting point, do chemical tests, hand in, and so on). Now, 0.1 g of a **liquid** is really very little. And if you assume that there are about 20 drops/mL for the average liquid and the average liquid has a density of 1 g/mL, you've got two whole microscale drops: one drop for the IR, one for the NMR, and for the boiling point . . . whoops!

Fortunately, most experimenters bend the scale so that you get close to 1 g (approximately 1 mL) when you prepare a liquid product. *That does not give you license to be sloppy!* Just be careful. Don't automatically put your liquid product in the biggest container you have. A *lot* of liquid can lose itself very quickly.

In this chapter I've included drawings of microscale equipment that I've had occasion to use, along with some discussion of the O-ring seals, conical vials, drying tubes, and so on. I've put full descriptions of certain microscale apparatus with the operations in which they're used. So Craig tubes show up with recrystallization; the Hickman still is with distillation.

UH-OH RINGS

Take a quick look at the floor of your organic chemistry laboratory. Now imagine yourself dropping a tiny, rubbery, brown or black O-ring on it. Uh-oh! The floors where I work are a common type of brown-, black-, and white-spotted terrazzo, and it's tough to find dropped O-rings. Make sure your hands are way over the bench top when you're monkeying with these things.

THE O-RING CAP SEAL

In the old days, you would put two glass joints together and, if you wanted a vacuum-tight seal, you'd use a little grease. These days, you're told to use a plastic cap and an O-ring to get a greaseless, high-tech seal. Maybe. Not all O-ring cap seal joints are created equal.

Skinny Apparatus

If you can drop the O-ring over the apparatus from the top, and it'll slide down to the bottom, such as fitting an air condenser to a conical vial; it's very easy to set up:

1. Get a conical vial (appropriate size) that'll fit on the bottom end (male joint) of the wide-bottom air condenser (Fig. 5.1).

2. Drop the O-ring over the top of this air condenser.

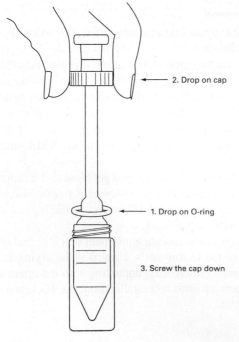

2. Drop on cap

1. Drop on O-ring

3. Screw the cap down

FIGURE 5.1 O-ring cap seal on skinny apparatus.

3. Drop the cap over the top of this air condenser.

4. Carefully screw the cap down onto the threads of the vial.

Not-So-Skinny Apparatus

Unfortunately, you *can't* drop an O-ring and cap over a water-jacketed condenser, Hickman still, and so on. What to do? Using the water-jacketed condenser as an example, go from the bottom up:

1. Get a conical vial (appropriate size) that'll fit on the end (male joint) of the condenser (Fig. 5.2).

2. Take the vial *off* the end.

3. Put a plastic cap onto the male joint and hold it there.

4. Push an O-ring up over the male joint onto the clear glass. The O-ring *should* hold the cap up. (Fingers may be necessary.)

5. Put the conical vial *back* on the male joint.

6. Carefully screw the cap down onto the threads of the vial.

Sizing Up the Situation

There have been several variations of the O-ring cap seal and you might still come across some of them:

1. *A 3- to 5-mL conical vial and ⊤ 14/10 joints.* The plastic caps for these sizes have holes cut in them that fit *extremely* tightly. They make a really good seal, and I suspect you would break the glass before *ever* pulling one of these

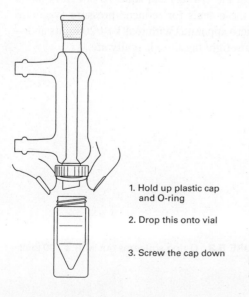

1. Hold up plastic cap and O-ring

2. Drop this onto vial

3. Screw the cap down

FIGURE 5.2 O-ring cap seal on husky apparatus.

babies apart. Once, when dismantling one of these joints, I caught the O-ring in the cap, and as I pulled the cap off the joint it sliced the O-ring in two. Watch it.

2. *A 0.3- to 1.0-mL conical vial and T 7/10 joint* (Fig. 5.3). The T joint rattles around in the hole in the plastic caps for this size. As a consequence, whenever you tighten these joints, the O-ring squeezes out from under the cap. Yes, the joint's tight, but gas-tight? You can pull the joints apart with modest effort, and, if the O-ring has squeezed out entirely, forget the cap as a stabilizing force.

3. *A 0.1-mL conical vial and T 5/5 joint.* For these sizes, again, the hole in the cap is snug around the male joint so that when you screw down the cap, the O-ring doesn't squeeze out. This joint also appears to be very tight.

Why I Don't Really Know How Vacuum-Tight These Seals Are

I'll break here and let Kenneth L. Williamson of *Macroscale and Microscale Organic Experiments* (D. C. Heath & Co., Lexington, Massachusetts, 1989, p. 102) take over. Here he is on the subject of microscale vacuum distillation assemblies: "On a truly micro scale (<10 mg), simple distillation is not practical because of mechanical losses."

That's clear enough. You lose lots of your microscale product with a reduced-pressure (vacuum) distillation. As far as I know, no microscale laboratory manual has anyone perform this. Yes, you might be asked to remove solvent under reduced pressure, but that's neither a reduced-pressure distillation nor uniquely microscale.

So, since nobody actually does any microscale reduced-pressure distillations, and since with certain sizes of equipment the O-rings can squeeze out from under the plastic caps, and since I don't use these seals for reduced-pressure (vacuum) distillation anyway (I switch to a semimicro apparatus with real T 14/20 joints and—gasp!—grease), I don't know how vacuum-tight these seals really are.

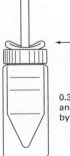

O-ring squeezes out

0.3- to 1.0-mL vial
and T 7/10 joint hidden
by plastic screw cap

FIGURE 5.3   O-ring squeezes out with T 7/10 joint.

THE COMICAL VIAL (THAT'S *CONICAL!*)

The **conical vial** (Fig. 5.4) is the round-bottom flask of the microscale set with considerably more hardware. When your glassware kit is brand-new, every vial has a plastic cap, O-ring, and plastic disk to match. After one semester, every vial *needs* this stuff. When you check in, make sure that every conical vial has a cap that fits and that you have at least one O-ring for each and a plastic disk that just fits inside this cap.

The Conical Vial as Vial

You would think that with an O-ring and a Teflon-faced plastic disk wedged under the cap, this would make a very tight seal. Think again. One of my students made *n*-hexylamine as a special project and packaged it in one of these vials; some of the amine leaked out, air got in, and the amine oxidized to a lovely brown color. You can still use them if you want, but be careful:

1. Check to see that the Teflon-faced plastic disk doesn't have any holes in it. Sometimes you pierce a disk with a syringe needle to add something to a setup. You use the disk as a puncturable seal. With a puncture or two, the seal won't be as good as with a fresh disk. Just to be sure, get a new one if your disk has had lots of punctures.

2. The Teflon side to the chemicals, please. Especially if the compound might attack the plastic disk. Teflon is *inert;* the plastic is *ert.*

3. The O-ring goes close to the cap. Otherwise, it can't press the disk down to the glass.

4. Now the cap. Screw this baby down and you should, *should,* mind you, have a very well-sealed vial.

Plastic cap

Silicone rubber O-ring

Teflon-faced plastic disk

⊺ ground joint in here

Standard glass thread out here for plastic cap

Graduations

Glass as thick as all get-out

FIGURE 5.4   The conical vial.

Laboratory gorillas please note:
Overtightening caps breaks the threads.
Thank you.

Even so, watch for leaks! Sometimes you're asked to perform an extraction in one of these vials, and you decide to shake the bejesus out of the vial to mix the two layers (see Chapter 16, "Extraction and Washing: Microscale"). After you clean your fingers, you *will* be more gentle when you repeat the experiment, won't you?

Packaging Oops

Did you get the weight of the product so that you could calculate the yield? ("What? *Calculation* in organic lab?") Fat lot of good that tight seal will do for you now that you have to undo it to weigh your product. And you'll lose a lot of your product on the walls of the vial. Nothing but problems, especially in microscale.

Tare to the Analytical Balance

You can avoid losing your product and save a bit of time if you weigh the empty vial *before* you put your product in it. Microscale quantities, unfortunately, may require a high-precision **analytical balance** rather than a **top-loading balance**, and you should be aware of at least one thing:

Analytical balances are very delicate creatures.
Only closed containers in the balance box!

Now, I don't know if you have a hanging pan analytical balance with dial-up weights or a fancy electronic model, and I don't care. Just keep your noxious organic products in closed containers, OK?

The Electronic Analytical Balance

1. Open your notebook and, on one line, write "Mass of empty container," and get ready to write in the mass. What? You didn't bring your notebook with you to record the weights? You were going to write the numbers down on an old paper towel? Go back and get your notebook.
2. Turn the balance on. If it's already on, zero the balance. See your instructor for the details.
3. Open the balance case door, and put the empty vial with cap and O-ring and plastic disk on the balance pan. Close the balance case door.
4. Wait for the number to stop jumping around. Even so, electronic digital balances can have a "bobble" of one digit in the least significant place. Write down the most stable number.
5. Take the vial out of the balance case. *Now* fill the vial with your product. Close it up. No open containers in the balance case.
6. Write this new number down. Subtract the numbers. You have the weight of your product and the minimum of loss.

"What if I have a **tare bar** or **tare knob**? Why not use this to zero the balance before putting in product? That way, I don't have to subtract—just get the weight of the product directly." Sure. Now take some of your product out of the vial to use in, say, another reaction. How much did you take? What? You say you already put it into the other reaction vial, and it's all set up and you can't take product out to weigh it? What? You say you think you'll lose too much product reweighing it in yet another vial before you commit it to your reaction? Did you record the tare of the product vial? No? One subtraction too many for you, huh?

But if you did record the weight of the empty vial, cap, and O-ring, all you need do is remove some product, close the vial up, weigh it again, and subtract *this* weight from the sum of the tare and original product mass.

"What if I have a top-loading balance without a cage—must I have only closed containers on it while weighing?" Yes. Cuts down on the mess. An open container on a balance pan is usually an irresistible invitation to load product into it while on the pan, spilling the product onto the pan.

"What about weighing paper?" What *about* weighing paper? You *always* fill or empty your containers *away from the balance pan,* and you always weigh your samples *in these closed containers,* so what about weighing papers?

Heating These Vials

Sources of heat vary according to what you have available. Some use a can-type heating mantle, only it's filled with sand. You raise or lower the vial in the sand pile to set the temperature you want. Others use a **crystallizing dish** on top of a **hot plate with magnetic stirrer** (Fig. 5.5). You put the vial flat on the crystallization dish and add sand. You can monitor the temperature of the vial roughly by sticking a thermometer in the sand.

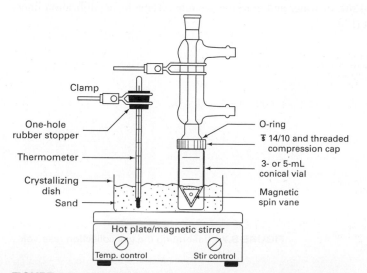

FIGURE 5.5 Hot bath for conical vial; just happens to be a reflux.

THE MICROSCALE DRYING TUBE

The **microscale drying tube** (Fig. 5.6) is just a right-angled glass tube with a ᛏ joint on one end. You put a small cotton plug in first and then load in the drying agent. Finish up with another cotton plug.

    If you let these sit around too long, the drying agent cakes up, and you could break the tube hacking the stuff out. It's actually cheaper to empty the tube after using it. If you can't hack that, soak the tube overnight (or so), and dissolve out the caked drying agent.

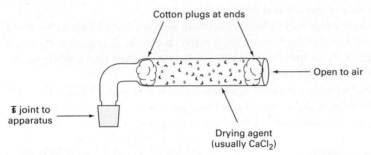

FIGURE 5.6 A microscale drying tube.

GAS COLLECTION APPARATUS

Several microscale kits have **capillary gas delivery tubes**. With this tube attached, you can collect gaseous products.

1. Put a cap on your **gas collection reservoir**. You can calibrate the tube by adding 2, 3, or 4 mL of water and marking the tube. Some have calibration lines. Check it out (Fig. 5.7).

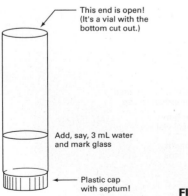

FIGURE 5.7 Calibrating the gas collection reservoir.

2. Get a 150-mL beaker and fill it with water to about 130 mL. A 100-mL beaker filled to 80 mL will work, but it'll be a bit tight.

3. Now fill the **gas collection reservoir** with water, put a finger over the open end, invert the tube, and stick the tube (and your fingers) under the water. Remove your finger. See that the gas collection reservoir remains filled.

4. Angle the straight end of the **capillary gas delivery tube** under and into the gas collection reservoir (Fig. 5.8). Believe it or not, the gas collection reservoir, sitting atop the capillary gas delivery tube, is fairly stable, so you can put this aside until you need it.

5. When you're ready, use a ring with a screen to support the beaker at the right height to connect the joint end of the capillary gas delivery tube to whatever you need to. Direct connection to a conical vial usually uses an O-ring cap seal; connection to a condenser may not (Fig. 5.9). Note that I've drawn the water connections to the condenser at an angle. Although the drawing is two-dimensional, your lab bench is not. You may have to experiment to find an arrangement that will accommodate the ring stand, hot plate, and other paraphernalia. Don't get locked into linear thinking.

Generating the Gas

Usually, you warm the conical vial somehow to have the reaction generate the gas you want to collect.

Do not stop heating once you start!
Water will get sucked back into the reaction mixture
as the apparatus cools!

Watch the flow of gas into the collection bottle. If you increase the heat and no more gas comes out, the reaction is probably done.

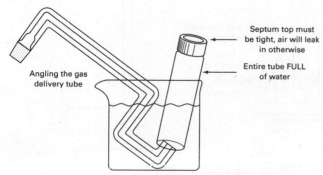

Angling the gas
delivery tube

Septum top must
be tight, air will leak
in otherwise

Entire tube FULL
of water

FIGURE 5.8 Setting up for gas collection.

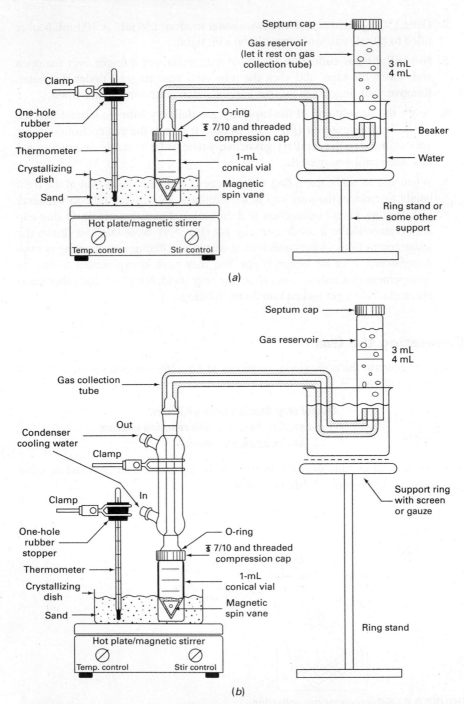

FIGURE 5.9 (a) Gas collection directly from a conical vial. (b) Gas collection from a condenser.

Remove the capillary gas delivery tube first.
Then remove the heat.

Of course, some gaseous reaction product, a few microliters perhaps, will still come off the mixture. Don't breathe this. And you *did* carry this out in a hood, no?

If your capillary gas delivery tube was connected to the top of a condenser (Fig. 5.9*b*), just pick up the beaker, collection vial, and capillary gas delivery tube, as a unit, off the condenser, and set it down. Then attend to the reaction.

If your capillary gas delivery tube was connected directly to a conical vial with an O-ring cap seal (Fig. 5.9*a*), you have to undo the seal on a hot vial (!) to lift the entire system off as a unit. And you can't let the vial cool with the capillary gas delivery tube under water, or water will be sucked back into the vial. The best I've been able to do is carefully lift the gas collection vial off the gas delivery tube (watch the water level in the beaker: If it's too low, the water in the collection vial will run out!) and then lower the beaker with the collection vial in it away from the capillary gas delivery tube. There's got to be a better way, but I don't know it yet. Ask your instructor.

Isolating the Product

With the **gas collection reservoir**, you have an O-ring cap seal with a Teflon-faced plastic insert. Pierce the insert with a hypodermic syringe and draw off some of the product. Usually, you put this gas through gas chromatography (see Chapter 30, "Gas Chromatography," for particulars) for analysis.

OTHER INTERESTING EQUIPMENT

- *Your thermometer should work.*
- *No glassware, even chipped.*

An early edition of this book illustrated some equipment specific to the State University of New York at Buffalo, since that's where I wrote it. Now it's a few years later, and I realize that you can't make a comprehensive list. SUNY, Buffalo, has an unusual pear-shaped distilling flask that I've not seen elsewhere. The University of Connecticut equipment list contains a Bobbitt filter clip that few other schools have picked up. So if you are disappointed that I don't have a list and drawing of every single piece of equipment in your drawer, I apologize. Only the most common organic lab equipment is covered here. Ask your instructor, "Whatizzit?" if you do not know. I used to assume (check previous editions if you doubt me) that you remembered Erlenmeyer flasks and beakers and such from freshman lab. But it's been *suggested* that I put drawings of even these beasties in this section anyway. OK by me. But I'll still be discussing the other apparatus as it comes up in the various techniques. This might force you to read this book before you start in the lab. So, check out the new and improved (yeah, right) Figure 6.1. Not all the mysterious doodads in your laboratory drawer are shown, but lots of old friends from your freshman chem lab are there, too. Just in case you forgot.

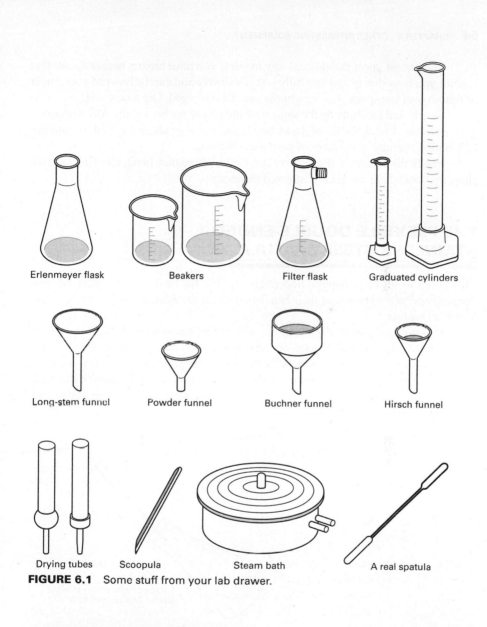

Erlenmeyer flask   Beakers   Filter flask   Graduated cylinders

Long-stem funnel   Powder funnel   Buchner funnel   Hirsch funnel

Drying tubes   Scoopula   Steam bath   A real spatula

FIGURE 6.1 Some stuff from your lab drawer.

FUNNELS, AND BEAKERS, AND FLASKS—OH MY!

While you are checking in, there are things about these pieces of nonjointware apparatus you should check out. Make sure that the pouring spouts on your beakers aren't cracked. This can happen when you nest the beakers—putting smaller ones inside larger ones—and the spout of the smaller beaker bonks up against the inside wall of the larger beaker and breaks.

The easiest, most paradoxical way to tell if you have broken beaker spouts that can cut you is to slowly and carefully—that's slowly and carefully—run your finger or thumb over the spout. Any roughness and it's damaged. Get a new one.

Slowly and carefully do the same with the rest of the beaker lip. Any damage—get a new one. Check the lip of flasks, spout and lip of graduated cylinders, and rim and angled opening at the stem of your glass funnels.

Check the bottom of the beakers and flasks where they bend, too. Cracks often show up here where the glass is stressed the most.

THE FLEXIBLE DOUBLE-ENDED STAINLESS STEEL SPATULA

This is not about the scoopula. Scoopulas are fine for freshman chemists, but real "organikers" would be lost at their benches without the flexible double-ended stainless steel spatula.

Do you have a solid product on the walls of an Erlenmeyer flask? Scrape it off the glass and onto the straight edge of one of the spatula paddles. Solid product stuck on the walls of a round-bottom flask? With one of the paddles bent—sort of curled, actually—you can easily scrape the solid down into a pile at the bottom of the flask, then lift it out with this kind of spatula. Try either of these common operations with a scoopula, and you will fail (Fig. 6.2).

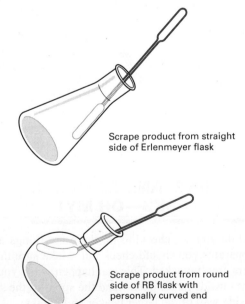

Scrape product from straight side of Erlenmeyer flask

Scrape product from round side of RB flask with personally curved end

FIGURE 6.2 Flexible double-ended stainless steel spatula at work.

I've seen two variants of the flexible double-ended stainless steel spatula. The first version is made of really heavy steel and the paddles can't bend. I believe the word *pointless* is preferred to *useless* here.

The second version has only one long paddle. The other end is a shorter paddle that comes down to a point. That pointed end is excellent for sticking the big rubber band on your lab goggles back under the molded slots when the band pulls out.

Transferring a Powdered Solid with the Spatula

1. Hold the bottle in one hand, and the spatula in the other.

2. Carefully, tilt the bottle until it is as horizontal as practical. Don't tilt the bottle so much that solid falls out! Jeezaloo! Who are you!

3. Move the spatula into the bottle, and pick up some solid on the paddle.

4. Before removing the spatula, tap the loaded paddle on the side of the bottle a few times. This knocks really loose powder back into the bottle, and reduces the likelihood of it falling off the spatula as you carry it.

PIPET TIPS

- *Solvents can blow out of a pipet and rubber bulb. Careful.*

- *Don't turn the pipet and bulb upside down! The liquid will contaminate the bulb.*

- *Pipets filter small amounts of liquid product that might disappear on a gravity filter.*

The Pasteur pipet (Fig. 7.1) is really handy for *all* scales of laboratory work, not just microscale, but it gets the most use in microscale. It usually comes in two sizes, a 9-in. (approximately 3-mL) pipet and a $5\frac{3}{4}$-in. (approximately 2-mL) pipet.

Although they look like eyedroppers, they aren't. The drop size of a Pasteur pipet is very different from that of a traditional eyedropper. I took the average mass of 20 drops of room temperature cyclohexanone and found that an eyedropper delivered about 0.0201 g and a Pasteur pipet delivered about 0.0137 g. The conventional wisdom in these things says there are 20 drops/mL, but these are eyedropper drops, not pipet drops.

PRE-PREPARING PASTEUR PIPETS

Figure 7.1 shows a rough calibration chart for both sizes.

Calibration

You should get an idea about how high in the pipet 0.5, 1.0, 1.5, and 2.0 mL of liquid are. Just draw measured amounts into the tube and look. I'd mark *one* pipet with all these

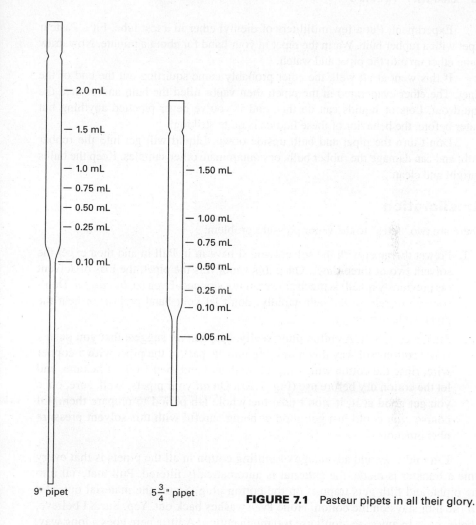

9" pipet	$5\frac{3}{4}$" pipet
— 2.0 mL	
— 1.5 mL	
— 1.0 mL	— 1.50 mL
— 0.75 mL	
— 0.50 mL	— 1.00 mL
— 0.25 mL	— 0.75 mL
	— 0.50 mL
— 0.10 mL	— 0.25 mL
	— 0.10 mL
	— 0.05 mL

FIGURE 7.1 Pasteur pipets in all their glory.

volumes and keep it as a reference. Marking your working pipets can get messy, especially when the markings dissolve into your product. And scratching a working pipet with a file is asking for trouble. Get the scratch wet, a little lateral pressure, and—snap—that's the end of the pipet.

Operation

I know I'm not supposed to have experiments in this book, but here's something I strongly suggest you carry out (translation: Do it!). Work in a hood. Wear goggles. Don't point the end of a pipet at anyone, yourself included.

Experiment: Put a few milliliters of diethyl ether in a test tube. Fit a Pasteur pipet with a rubber bulb. Warm the pipet in your hand for about a minute. Now draw some ether up into the pipet and watch.

If this went at all well, the ether probably came squirting out the end of the pipet. The ether evaporated in the pipet; then vapor filled the bulb and pushed the liquid out. Lots of liquids can do this, and if you've never pipetted anything but water before, the behavior of these liquids is quite striking.

Don't turn the pipet and bulb upside down. Liquid will get into the rubber bulb, and can damage the rubber bulb, or contaminate other samples. Keep the bulbs upright and clean.

Amelioration

There are two "fixes" to the vapor pressure problem:

1. Prewet the pipet with the solvent you'll have in it. Pull in and then expel the solvent two or three times. Once this vapor fills the pipet, the loss of solvent (as previously noted) is much slower, if not stopped. Even so, *be careful*. Don't squeeze or release the bulb rapidly; don't let your hand preheat or heat the pipet while handling it.

2. Stick a cork in it. A cotton plug, really. Mayo *et al.* suggest that you push a small cotton ball way down into the narrow part of the pipet with a copper wire, rinse the cotton with 1 mL of methanol and then 1 mL of hexane, and let the cotton dry before use (Fig. 7.2*a*). On *all* your pipets. Well, sure, once you get good at it, it won't take the whole lab period to prepare them. Of course, you could just get good at being careful with this solvent pressure phenomenon.

I'm told a second advantage of putting cotton in all the pipets is that every time a transfer is made, the material is automatically filtered. Pull material into the pipet, and filth gets trapped on the cotton plug. Squirt the material out, and *all* the filth stays on the cotton. None *ever* washes back out. Yep. Sure. I believe, I believe . . . In any case, don't use too much cotton! A little here goes a long way.

PIPET CUTTING

From time to time you might need to shorten the long tip of the Pasteur pipet so that the narrow portion is, say, 2–3 cm rather than the 8 cm or so of the 9-in. pipet.

1. Get a sharp triangular file or, better yet, a sharp glass scorer.

2. Support the narrow part of the tip on a raised solid surface such as a ring-stand plate.

3. Carefully draw the cutter back toward you while you rotate the pipet away from you. Don't press down too much; you'll crush the tip.

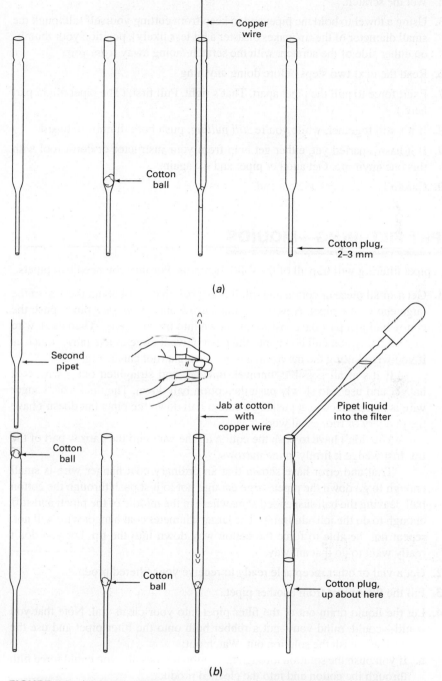

FIGURE 7.2   (*a*) Pasteur filter pipet—Mayo. (*b*) Pasteur filter pipet—Zubrick.

4. Wet the scratch.

5. Using a towel to hold the pipet and to keep from cutting yourself (although the small diameter of the tip makes disaster a bit less likely), position your thumbs on either side of the scratch, with the scratch facing away from you.

6. Read the next two steps before doing anything.

7. Exert force to pull the pipet apart. That's right. Pull first. (The pipet might part here.)

8. If it's still together, while you're *still pulling,* push both thumbs outward.

9. If it hasn't parted yet, either get help from your instructor or don't fool with this one anymore. Get another pipet and try again.

10. Carefully fire-polish the cut end.

PIPET FILTERING—LIQUIDS

Real pipet filtering will trap all of the solid materials. For this you need two pipets.

1. Get a small piece of cotton and roll it into a ball that's just about the size of the large end of the pipet. Now, using the narrow end of *another* pipet, push the cotton ball just past the constriction, down into the first pipet. Then use a wire to push the cotton ball down into the pipet. Give it three to five jabs—much as if you were killing the nerves in a tooth during a root canal (Fig. 7.2*b*).

   If it's at all possible, unravel, unbend, and straighten out a wire coat hanger and use it to slowly push the cotton ball down. The thick coat hanger wire is more effective at pushing a cotton ball down the pipet but doesn't have the imagery of root canal work.

   You don't have to push the cotton all the way into the narrow part of the tip. Just wedge it firmly in the narrows.

   Trial and error have shown that an ordinary coat hanger wire is small enough to go down the pipet; large enough not to just push through the cotton ball, leaving the ball suspended somewhere in the middle of the pipet; and stiff enough to do the job admirably. The larger-diameter coat hanger wire will not, repeat not, be able to force the cotton way down into the tip, but you don't really want to do that anyway.

2. Get a vial or other receptacle ready to receive your filtered product.

3. Fill the filter pipet from another pipet.

4. Let the liquid drain out of the filter pipet into your clean vial. Note that you could—could, mind you—put a rubber bulb onto the filter pipet and use the bulb to help push the solution out. Watch out:

   a. If you push the solution through the cotton too rapidly, you could force filth through the cotton and into the cleaned product.

   b. Don't *ever* let up on the rubber bulb once you start. You'll suck air and possibly filth off the cotton back into the pipet. In a few cases, where people didn't jam the cotton into the tip, but had the cotton as a large ball wedged in the narrowing part, the cotton would be sucked back into the pipet, and the entire solution, filth and all, would run right out of the pipet.

c. You could—this is rare—blow the cotton jammed in the tip right out.

PIPET FILTERING—SOLIDS

After a recrystallization, you usually collect the new crystals by suction on a Buchner funnel (see Chapter 13, "Recrystallization"). For microscale quantities, you may have to use a Hirsch funnel—a tiny Buchner funnel with sloping sides and a flat, porous plate (see Chapter 6, "Other Interesting Equipment"). Or you might need the high-tech power of Craig tubes (see Chapter 14, "Recrystallization: Microscale"). Or you might be able to get away with a Pasteur pipet (Fig. 7.3).

1. Get the vial or tube containing the ice-cold crystalline mixture and stir it a bit with a Mayo type cotton-plugged pipet—the cotton plug pushed right to the end of the tip.

2. Slowly—there's that word again—push air out of the pipet while you bring the tip to the bottom of the container.

3. Slowly—there's that word yet again—draw solvent into the pipet, leaving crystals behind.

Now you might want to pack the crystals down a bit by rapping the container on a hard surface. Then repeat the filtration above.

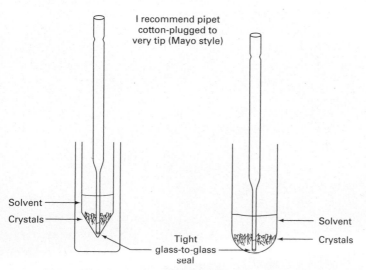

FIGURE 7.3 Pasteur pipets to filter solids.

You can wash these crystals with a few drops of cold, fresh solvent and then remove the solvent as many times as is necessary. Eventually, however, you'll have to take the crystals out of the tube and let them dry.

If the tube with crystals has a flattened or even a slightly rounded bottom, you might be able to get away with using an unplugged pipet. The square end of the pipet can fit even a test tube bottom fairly closely, letting solvent in and keeping crystals out. With conical vials, there's more trouble getting a close glass-to-glass contact (not impossible, though), and it might be easier to stick to the Mayo-type plugged pipet (Fig. 7.3).

In all cases, make sure you keep the crystallization tube cold and that you wait, even the small length of time it takes, for the pipet tip to cool as well. You don't want the pipet tip warming the solution.

SYRINGES, NEEDLES, AND SEPTA

- *Don't fool with these things and stick yourself or another.*
- *Don't use the M.D. injection technique; these are not darts.*

Often, you have to handle samples by syringe. There are about 25 rules for handling syringes. The first five are:

1. These are dangerous—watch it.
2. Watch it—don't stick yourself.
3. Don't stick anyone else.
4. Be careful.
5. Don't fool around with these things.

The other 20 rules—well, you get the idea. If that's not enough, some states require *extremely* tight control of syringes and needles. They might be kept under double lock and key in a locked cabinet in a locked room and require extensive sign-out *and* sign-in procedures. And if you do get stuck, bring yourself and the syringe to the instructor immediately.

You'll probably use a glass or plastic syringe with a **Luer tip** (Fig. 8.1). This is a standard-shape tip that can fit various pieces of apparatus, including the syringe needle. All you do is push the tip into the receptacle. Don't confuse this with the **Luer-Lok**, a special collar placed around the Luer tip designed to hold, well, *lock* the syringe barrel onto the apparatus, including needles (Fig. 8.2). You push the lock onto the receptacle and, with a twist, lock the syringe barrel to it. You'll have to *untwist* this beast to release the syringe barrel.

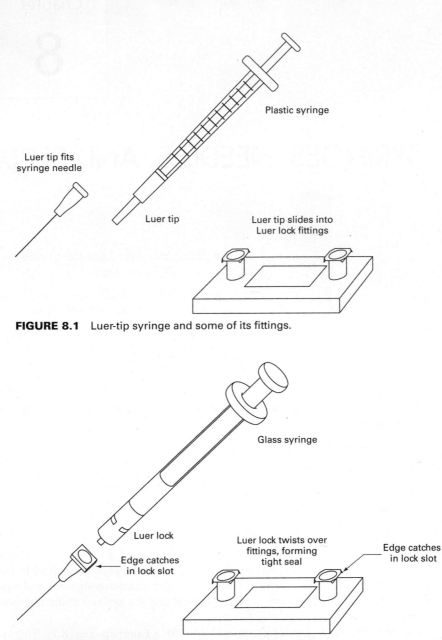

Plastic syringe

Luer tip fits
syringe needle

Luer tip

Luer tip slides into
Luer lock fittings

FIGURE 8.1 Luer-tip syringe and some of its fittings.

Glass syringe

Luer lock

Edge catches
in lock slot

Luer lock twists over
fittings, forming
tight seal

Edge catches
in lock slot

FIGURE 8.2 Luer-Lok syringe and some of its fittings.

The biggest mistake—one that I still make—is losing my patience when filling a syringe with a needle attached. I always tend to pull the plunger quickly, and I wind up sucking air into the barrel. Load the syringe slowly. If you do get air in the barrel, push the liquid back out and try again.

Injecting a sample through a Teflon-lined plastic disk, or a rubber septum, is fairly straightforward. Don't get funny and try slinging the needle—medical school-style—through the disk. Use two hands if you have to: one to hold the barrel and the other to keep the plunger from moving. Slowly push the plunger into the barrel to deliver the liquid. Microliter syringe handling and injection through a septum is also discussed in Chapter 30, "Gas Chromatography."

THE RUBBER SEPTUM

If you're using syringes, you should know about the **rubber septum**. You put the smaller end into the opening you're going to inject into, and then roll the larger end over the *outside* of the opening (Fig. 8.3). You might use a length of copper wire to keep the rubber septum from coming out of the opening, especially if there's any chance of a pressure buildup inside the apparatus.

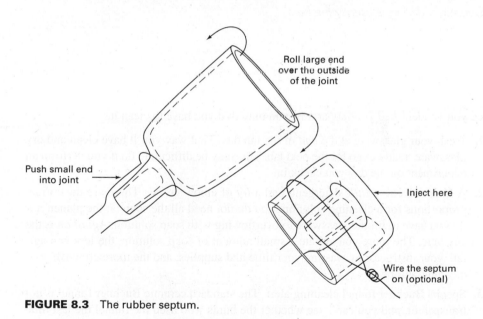

Roll large end over the outside of the joint

Push small end into joint

Inject here

Wire the septum on (optional)

FIGURE 8.3 The rubber septum.

CLEAN AND DRY

- *Use elbow grease, too.*
- *Don't dry stuff that'll have water in it.*
- *Overnight drying is usually the best.*

Once you've identified your apparatus, you may find you have to clean it.

1. Wash your glassware at the *end* of the lab day. That way you'll have clean and dry glassware, ready to go for the next lab. This may be difficult to do if you perform an experiment on the day you check in.

2. A *little* solvent, a *little* detergent, and a *lot* of elbow grease. These are the correct proportions for a cleaning solution. You do not need all the soap on the planet, nor do you have to fill the glassware to overflowing with soap solution. *Agitation* is the key here. The more you agitate a small amount of soap solution, the less you agitate your instructor by wasting your time and supplies, and the more effective your cleaning will be.

3. Special Buchner funnel cleaning alert: The standard ceramic Buchner funnel is not transparent, and you can't see whether the bums who used the funnel the last time to collect a highly colored product cleaned the funnel properly. The first time you Buchner-filter crystals from an alcohol solution, the colored impurity dissolves and bleeds up into your previously clean crystals, and you may have to redo your entire experiment. I'd rinse the Buchner funnel with a bit of hot ethanol before I used it, just for insurance.

DRYING YOUR GLASSWARE WHEN YOU DON'T NEED TO

"It's late. Why haven't you started the experiment yet?"

"I washed all my glassware and spent half an hour drying it."

"What technique are we doing?"

"Steam distillation."

"Steam goes through the entire setup, does it?"

A nodding head responds.

"What's condensed steam?"

"Water . . ."

There are all sorts of variations, but they boil down to this: You've taken all this time to dry your glassware only to put water in it. Writers of lab manuals are very tricky about this. Perhaps they say you'll be using *steam*. Or maybe *5% aqueous* sodium bicarbonate solution. Or even that a by product of your reaction is H_2O. Condense *steam* and you get *what*? An *aqueous* solution has *what* for a solvent? H_2O is *what*?

Look for sources of water other than plain water. If a "water-and" mixture is going to be in the equipment anyway, drying to perfection is silly.

DRYING YOUR GLASSWARE WHEN YOU DO NEED TO

If you wash your glassware *before* you quit for the day, the next time you need it, it'll be clean and dry. There are only a few reactions you might do that need superclean, superdry apparatus, and you should be given special instructions when that's necessary. (In their book *Experimental Organic Chemistry,* 2nd edition, McGraw-Hill, 1986, authors H. D. Durst and G. W. Gokel claim that glassware dried overnight is dry enough for the Grignard reaction—an *extremely* moisture-sensitive reaction—and that flame-drying can be avoided unless the laboratory atmosphere is extremely humid.)

Don't use the compressed air from the compressed air lines in the lab for drying anything. These systems are full of dirt, oil, and moisture from the pumps, and they will get your equipment dirtier than before you washed it.

Yes, there are a few quick ways of drying glassware in case of emergency. You can rinse *very wet* glassware with a small amount of acetone, *drain the glassware very well,* and put the glassware in a drying oven (about 100°C) for a short spell. The acetone washes the water off the glassware very well (the two liquids are **miscible**, that is, they mix in all proportions), and also the liquid left behind is acetone-rich and evaporates faster than water.

If you have one of those "kits" to store your glassware in, make really sure all the acetone is gone. The plastic liners of some of these kits can dissolve in acetone—and there you'll be next time, chipping your glassware out of your kit.

Chapter 10

DRYING AGENTS

- *Add small amounts until the job is done.*
- *If there's dry drying agent in the flask, the liquid must be dry.*

When you've prepared a liquid product, you must dry the liquid before you finally distill and package it by treating the liquid with a **drying agent**. Drying agents are usually certain anhydrous salts that combine with the water in your product and hold it as **water of crystallization**. When all of the water in your sample is tied up with the salt, you gravity-filter the mixture. The dried liquid passes through the filter paper, and the **hydrated salt** stays behind.

TYPICAL DRYING AGENTS

1. *Anhydrous calcium chloride.* This is a very popular drying agent, inexpensive and rapid, but of late I've become disappointed in its performance. It seems that the calcium chloride powders a bit from storage and abuse, and this **calcium chloride dust** go right through the filter paper with the liquid. So a caution: If you must use anhydrous calcium chloride, be sure it is granular. There are 4–8-mesh pellets that work fairly well. Avoid powdered calcium chloride, or granular anhydrous calcium chloride that's been around long enough to become pulverized. And don't add to the problem by leaving the lid off the jar of the drying agent; that's the abuse I was talking about.

   Anhydrous calcium chloride tends to form *alcohols of crystallization,* so you really can't use it to dry alcohols.

2. *Anhydrous sodium carbonate and anhydrous potassium carbonate.* These are useful drying agents that are basically **basic**. As they dry your organic compound, any carbonate that dissolves in the tiny amounts of water in your sample can neutralize any tiny amounts of acid that may be left in the liquid. If your product is *supposed* to be acidic (in contrast to being *contaminated* with acid), you should avoid these drying agents.

3. *Anhydrous sodium sulfate.* This is a neutral drying agent that has a high capacity for water, so you can remove huge quantities of water from very wet samples. It's a little slow, and not too efficient, but it is cheap, and it forms a **granular** hydrate from which you can easily decant your dried liquid. It loses water above 32°C, so you might have to pay attention to the solution temperature.

4. *Anhydrous magnesium sulfate.* In my opinion, anhydrous (anh.) $MgSO_4$ is about the best all-around drying agent. It has a drawback, though. Since it's a fine powder, lots of your product can become trapped on the surface. *This is not the same as water of crystallization.* The product is *only on the surface, not inside the crystal structure.* You can wash your product out of the powder with a little fresh solvent.

5. *Drierite.* Drierite, a commercially available brand of anhydrous calcium sulfate, has been around a long time and is a popular drying agent. You can put it in liquids and dry them or pack a drying tube with them to keep the moisture in the air from getting into the reaction setup. But be warned: There is also **blue Drierite.** This has an indicator, a cobalt salt, that is *blue when dry, pink when wet.* Now, you can easily tell when the drying agent is no good. Just look at it. Unfortunately, this stuff is not cheap, so don't fill your entire drying tube with it just because it'll look pretty. Use a small amount mixed with white Drierite, and when the blue pieces turn pink, change the entire charge in your drying tube. You can take a chance using blue Drierite to dry a liquid directly. Sometimes the cobalt compound dissolves in your product. Then you have to clean and dry your product all over again.

USING A DRYING AGENT

1. Put the liquid or solution to be dried into an Erlenmeyer flask.

2. Add *small* amounts of drying agent and swirl the liquid. Each time you swirl and stop, see if the powered drying agent is taking some time to settle down, and looks like, well, dry drying agent. If the drying agent in the liquid is dry, then the liquid is dry. Larger granulated drying agents might not behave so nicely, and fall rapidly, wet or dry. If you can see that the liquid is no longer cloudy, the water is gone and the liquid is dry.

3. Add just a bit more drying agent and swirl one final time.

4. Gravity-filter through filter paper (see Figure 13.3), or Pasteur pipet (see Figure 7.2).

5. If you've used a carrier solvent, then evaporate or distill it off, whichever is appropriate. Then you'll have your clean, dry product.

FOLLOWING DIRECTIONS AND LOSING PRODUCT ANYWAY

"Add 5 g of anhydrous magnesium sulfate to dry the product." Suppose your yield of product is lower than that in the book. Too much drying agent and not enough product—zap! It's all sucked onto the surface of the drying agent. Bye-bye product. Bye-bye grade.

Always add the drying agent slowly to the product in small amounts.

Now about those small amounts of product (usually liquids):

1. Dissolve your product in a *low-boiling-point solvent.* Maybe ether or hexane or the like. Now dry this whole solution, and gravity-filter. Remove the solvent carefully. Hoo-ha! Dried product.

2. Use chunky dehydrating agents such as anhydrous calcium sulfate (Drierite). Chunky drying agents have a much smaller surface area, so not as much of the product is absorbed.

DRYING AGENTS: MICROSCALE

Using drying agents in microscale work follows the same rules as drying on larger scales. Chunky Drierite becomes a problem, and you may have to chop it up a bit (good luck!) before you use it. And, as ever, beware directions that command you to fill a cotton-plugged pipet (see Chapter 7, "Pipet Tips") with exactly 2.76 g of drying agent: Again, this is a *typical* amount of drying agent to be used to dry the *typical* amount of product. The one nice thing about drying your liquids this way is that you can also filter impurities out of the liquid at the same time (see Figure 7.2b).

And remember, at microscale, it's more important than ever to consider washing the drying agent with fresh solvent to recover as much of your product as possible.

DRYING IN STAGES: THE CAPACITY AND EFFICIENCY OF DRYING AGENTS

Actually, using drying agents can be a bit more complicated than I've shown, especially if you have to choose the agent on your own. Fact is, if you dry an organic liquid with anhydrous sodium sulfate alone, you could wind up with water still present in the organic liquid, even if it looks dry (not cloudy). Sodium sulfate has a high capacity—it binds a lot of water per gram—but a low efficiency—it leaves water in the organic liquid. So why use it? Sodium sulfate is cheap. Coupled with its high capacity, it is ideal for "predrying" liquids; in the second drying stage, you could use a smaller amount of a lower-capacity, higher-efficiency drying agent.

EXERCISES

1. Phosphorous pentoxide is a heck of a drying agent. In a sealed enclosure, after some time there is no measurable amount of water in the air above the pentoxide. It is also entirely unsuitable for drying many liquids or solutions you would come across in your organic laboratory. What's with that?

2. Anhydrous potassium carbonate is useful for drying amines. Acids, not so much. Why?

3. You're drying a product with anhydrous magnesium sulfate. You're not entirely sure if you have added enough. Discuss the phrase "If you see dry drying agent in the flask, it's dry" in this context.

ON PRODUCTS

- *Samples clean and dry.*
- *Properly labeled vials, eh?*

The fastest way to lose points is to hand in messy samples. Lots of things can happen to foul up your product. The following are unforgivable sins! Repent and avoid!

SOLID PRODUCT PROBLEMS

1. ***Trash in the sample.*** Redissolve the sample, gravity-filter, and then evaporate the solvent.

2. ***Wet solids.*** Press out on filter paper, break up, and let dry. The solid shouldn't stick to the sides of the sample vial. Tacky!

3. ***Extremely wet solids (solid floating in water).*** Set up a gravity filtration (see Chapter 13, "Recrystallization") and filter the liquid off the solid. Remove the filter paper cone with your solid product, open it up, and leave it to dry. Or remove the solid and dry it on fresh filter paper as above. Use lots of care, though. You don't want filter paper fibers trapped in your solid.

LIQUID PRODUCT PROBLEMS

1. ***Water in the sample.*** This shows up as droplets or as a layer of water on the top or the bottom of the vial, or *the sample is cloudy*. Dry the sample with a drying agent (see Chapter 10, "Drying Agents"), and gravity-filter into a clean, dry vial.

2. *Trash floating in the sample.* For that matter, it could be on the bottom, lying there. Gravity-filter into a clean, dry vial.

3. *Water in the sample when you don't have a lot of sample.* Since solid drying agents can absorb lots of liquid, what can you do if you have a tiny amount of product to be dried? Add some solvent that has a low boiling point. It must dissolve your product. Now you have a lot of liquid to dry, and *if a little gets lost, it is not all product.* Remove this solvent after you've dried the solution. Be careful if the solvent is flammable. *No flames!*

THE SAMPLE VIAL

Sad to say, an attractive package can sell an inferior product. So why not sell yours? Dress it up in a **neat new label**. Put on:

1. *Your name.* Just in case the sample gets lost on the way to camp.

2. *Product name.* So everyone will know what is in the vial. What does "Product from part C" mean to you? Nothing? Funny, it doesn't mean anything to instructors either.

3. *Melting point (solids only).* This is a range, like "mp 96–98°C" (see Chapter 12, "The Melting-Point Experiment").

4. *Boiling point (liquids only).* This is a range, like "bp 96–98°C" (see Chapter 19, "Distillation").

5. *Yield.* If you weigh the empty vial and cap, you have the **tare**. Now add your product and weigh the full vial. Subtract the *tare* from this *gross weight* to get the **net weight** (yield, in grams) of your product.

6. *Percent yield.* Calculate the percent yield (see Chapter 2, "Keeping a Notebook"), and put it on the label.

You may be asked for more data, but the things listed above are a good start down the road to good technique.

P.S. Gummed labels can fall off vials, and pencil will smear. *Always use waterproof ink!* And a piece of transparent tape over the label will keep it on.

HOLD IT! DON'T TOUCH THAT VIAL

Welcome to "You Bet Your Grade." The secret word is **dissolve**. Say it slowly as you watch the cap liner in some vials dissolve into your nice, clean product and turn it all goopy. This can happen. A good way to prevent it is to cover the vial with aluminum foil before you put the cap on. Just make sure the product does not react with aluminum. Discuss this procedure at length with your instructor.

Chapter 12

THE MELTING-POINT EXPERIMENT

- *Never remelt samples.*
- *Tiny amounts. If you can just see it, you can see it melt.*

A **melting point** is the temperature at which the first crystal just starts to melt until the temperature at which the last crystal just disappears. Thus the melting point (abbreviated mp) is actually a **melting range**. You should report it as such, even though it is *called* a melting point, for example, mp 147–149°C.

People always read the phrase as melting *point* and never as *melting* point. There is this uncontrollable, driving urge to report one number. No matter how much I've screamed and shouted at people *not* to report one number, they almost always do. It's probably because many places list only one number, the upper limit.

Generally, melting points are taken for two reasons.

1. *Determination of purity.* If you take a melting point of your compound and it starts melting at 60°C and doesn't finish until 180°C, you might suspect something is wrong. A melting range *greater than 2°C* usually indicates an impure compound. (As with all rules, there are exceptions. There aren't many to this one, though.)

2. *Identification of unknowns.*
   a. If you have an unknown solid, take a melting point. Many books (ask your instructor) contain tables of melting points and lists of compounds that may have particular melting points. One of them may be your unknown. You may have 123 compounds to choose from. A little difficult, but that's not all the compounds in the world. Who knows? Give it a try. If nothing else, you know the melting point.

b. Take your unknown, and mix it *thoroughly* with some chemical you think might be your unknown. You might not get a sample of it, but you can ask. Shows you know something. Then,

(1) If the mixture melts at a *lower* temperature, over a *broad range,* your unknown is not the same compound.

(2) If the mixture melts at the *same temperature, same range,* it's a good bet it's the *same compound.* Try another run, though, with a different ratio of your unknown and this compound just to be sure. A *lower* melting point with a *sharp range* would be a special case called a **eutectic mixture**, and you, with all the other troubles in lab, just might accidentally hit it. On lab quizzes, this procedure is called

"Taking a mixed melting point."

Actually, "taking a *mixture* melting point," the melting point of a mixture, is more correct. But I have seen this expressed both ways.

SAMPLE PREPARATION

You usually take melting points in thin, closed-end tubes called **capillary tubes**. They are also called **melting-point tubes** or even **melting-point capillaries**. The terms are interchangeable, and I'll use all three.

Sometimes you may get a supply of tubes that are open on *both ends!* You don't just use these as is. Light a burner and, before you start, close off one end. Otherwise your sample will fall out of the tube (see "Closing Off Melting-Point Tubes" later in this chapter).

Take melting points only on *dry, solid* substances, *never* on liquids or solutions of solids *in* liquids or on wet or even damp solids. Only on dry solids!

To help dry damp solids, place the damp solid on a piece of filter paper and fold the paper around the solid. Press. Repeat until the paper doesn't get wet. Yes, you may have to use fresh pieces of paper. Try not to get filter paper fibers in the sample, OK?

Occasionally, you may be tempted to dry solid samples in an oven. *Don't*— unless you are specifically instructed to. I know some students who have decomposed their products in ovens and under heat lamps. With the time they save quickly decomposing their product, they can repeat the entire experiment.

Loading the Melting-Point Tube

Place a small amount of *dry* solid on a new filter paper (Fig. 12.1). Thrust the open end of the capillary tube into the middle of the pile of material. Some solid should be trapped in the tube. *Carefully* turn the tube over, taking care not to let the solid fall back out of the tube, until the closed end is down. Remove any solid sticking to the outside. Now, the solid must be packed down. If you have trouble with gouging the filter paper, you might put the solid on a watch glass instead of filter paper.

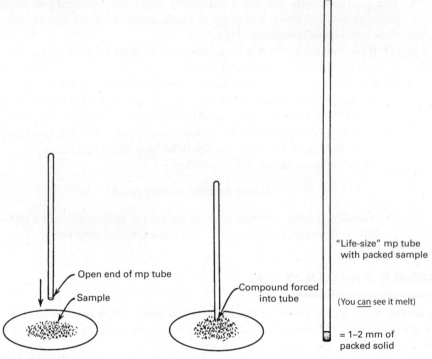

Open end of mp tube

Sample

Compound forced
into tube

"Life-size" mp tube
with packed sample

(You can see it melt)

= 1–2 mm of
packed solid

FIGURE 12.1   Loading a melting-point tube.

 Traditionally, the capillary tube, turned upright with the *open end up,* is stroked with a file or trapped on the bench top. Unless they are done *carefully,* these operations *may break the tube.* A safer method is to drop the tube, *closed end down,* through a length of glass tubing. You can even use your condenser or distilling column for this purpose. When the capillary strikes the bench top, the compound will be forced into the closed end. You may have to do this several times. If there is not enough material in the mp tube, thrust the open end of the tube into the mound of material and pack it down again. Use your own judgment; consult your instructor.

Use the smallest amount of material that can be seen to melt.

Closing Off Melting-Point Tubes

If you have melting-point tubes that are *open at both ends* and you try to take a melting point with one, it should come as no surprise when your compound falls out of the tube. You'll have to *close off one end* to keep your sample from falling

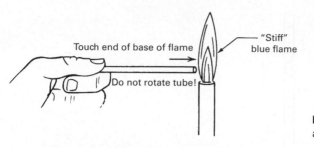

Touch end of base of flame

"Stiff"
blue flame

Do not rotate tube!

FIGURE 12.2   Closing off
an mp tube with a flame.

out (Fig. 12.2). So light a burner, and get a "stiff" small blue flame. Slowly touch
the end of the tube to the side of the flame, and hold it there. You should get a yellow sodium flame, and the tube will close up. There is no need to rotate the tube.
And remember, *touch—just touch*—the edge of the flame, and hold the tube there.
Don't feel you have to push the tube way into the flame.

MELTING-POINT HINTS

1. Use only the smallest amount that you can see melt. Larger samples will heat
   unevenly.

2. Pack down the material as much as you can. Left loose, the stuff will heat
   unevenly.

3. Never remelt any sample. It may undergo nasty chemical changes such as
   oxidation, rearrangement, or decomposition.

4. Make up more than one sample. One is easy, two is easier. If something goes wrong
   with one, you have another. Duplicate and even triplicate runs are common.

THE MEL-TEMP APPARATUS

The Mel-Temp apparatus (Fig. 12.3) substitutes for the Thiele tube or open beaker
and hot oil methods (see "Using the Thiele Tube" later in this chapter). Before you
use the apparatus, there are a few things you should look for.

1. *Line cord.* Brings AC power to the unit. Should be plugged into a live wall
   socket. (See J. E. Leonard and L. E. Mohrmann, *J. Chem. Educ.,* **57**, 119,
   1980) for a modification in the wiring of older units to make them less lethal.
   It seems that, even with the three-prong plug, there can still be a shock hazard.
   Make sure your instructor knows about this!)

2. *On–off switch.* Turns the unit on or off.

3. *Fuse.* Provides electrical protection for the unit.

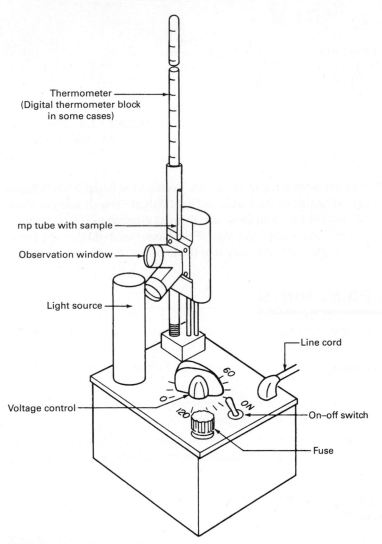

Thermometer
(Digital thermometer block
in some cases)

mp tube with sample

Observation window

Light source

Line cord

Voltage control

On–off switch

120

Fuse

FIGURE 12.3   The Mel-Temp apparatus.

4. *Voltage control.* Controls the *rate* of heating, *not the temperature*! The higher the setting, the faster the temperature rises.

5. *Light source.* Provides illumination for samples.

6. *Eyepiece.* Magnifies the sample (Fig. 12.4).

7. *Thermometer.* Gives the temperature of the sample, and upsets the digestion when you're not careful and you snap it off in the holder.

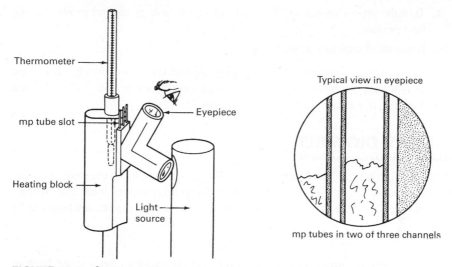

FIGURE 12.4   Close-up of the viewing system for the Mel-Temp apparatus.

Operation of the Mel-Temp Apparatus

1. *Imagine yourself getting burned if you're not careful.* Never assume that the unit is cold. Read the temperature, and wait for the heating block to cool if the temperature is fewer than 20°C below the melting point of your compound. You'll have to wait for the block to cool to room temperature if you have no idea of the approximate melting point of your compound.

2. Place the loaded mp tube in one of the three channels in the opening at the top of the unit (Fig. 12.4).

3. Set the voltage control to zero if necessary. There are discourteous folk who do not reset the control when they finish using the equipment.

4. Turn the on–off switch to ON. The light source should illuminate the sample. If not, call for help.

5. Now science turns into art. Set the *voltage control* to *any* convenient setting. The point is to get up to *within 20°C* of the *supposed* melting point. Yep, that's right. If you have no idea what the melting point is, it may require several runs as you keep skipping past the points with a temperature rise of 5–10°C per minute. A convenient setting is *40*. This is just a suggestion, not an article of faith.

6. After you've melted a sample, *throw it away!*

7. Once you have an idea of the melting point (or looked it up, or were told), *get a fresh sample,* and bring the temperature up quickly at about *5–10°C per minute to within 20°C* of this approximate melting point. Then, turn down the *voltage control* to get a *2°C per minute rise.* Patience!

8. When the first crystals *just start to melt,* record the temperature. When the *last crystal just disappears,* record the temperature. If both points appear to be the same, either the sample is extremely pure or the temperature rise was *too fast.*

9. Turn the on–off switch to OFF. You can set the voltage control to zero for the next person.

10. Remove all capillary tubes.

Never use a wet rag or sponge to quickly cool off the heating block. This might permanently warp the block. You can use a cold metal block to cool it if you're in a hurry. Careful. If you slip, you may burn yourself.

THE SRS DIGIMELT

The Mel-Temp for the 21st century. Physically, they have a lot in common, and you should probably read the section on the Mel-Temp apparatus as well. The instructions on how to use the unit are right on the front, so what I need to say here can be shorter (Fig. 12.5).

If you are the first one to use the unit:

1. Make sure the line cord is plugged into both the wall socket and the back of the unit.

2. The on–off switch is at the back of the unit next to the line cord. When you turn the unit on, the LED segments go in a circle, display "srs," show a software version number (my unit was 2.10), then display the temperature of the heating block. The "cooling" LED on the display panel lights up, not necessarily because the unit is cooling, but because it is not heating up.

3. Press START TEMP and use the arrow keys to set that temperature, press RAMP RATE and use the arrow keys to set a heating rate (2°C/min, eh?), and press STOP TEMP to set the maximum temperature you want to subject your sample to.

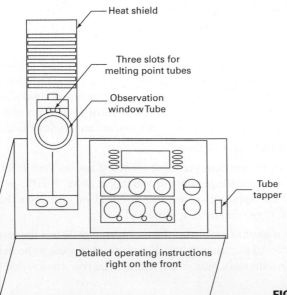

Heat shield

Three slots for melting point tubes

Observation window Tube

Tube tapper

Detailed operating instructions right on the front

FIGURE 12.5 The SRS DigiMelt.

4. Press the START/STOP button once, and the display goes back to reading the temperature of the heating block.

5. If necessary, take your correctly prepared melting point tube, put it in one of the spaces in the TUBE TAPPER block, and press the TUBE TAPPER button. Your sample will be vibrated and packed down.

6. Put the tube in one of the melting point slots at the top of the heating block. Make sure you can see your sample.

7. Press the START/STOP button and the unit will begin heating the block up to your START TEMP. The "preheat" LED comes on. When it starts to reach the START TEMP, the heating rate slows and then stops. It'll be within about $\pm 0.2°C$ of the START TEMP.

8. Press the START/STOP button again, get your finger over the blue "1" button, and watch your sample. When it just starts to melt, press that blue button. The unit has just recorded that temperature. Wait until the sample melts completely, and press the blue button again. The unit has just recorded that temperature. The "data" LED will be lit, indicating that there is data stored in the unit.

9. At the end of the run, when the block has reached your STOP TEMP, the unit stops heating the block and it starts to cool back to room temperature.

10. Each blue button ("1," "2," and "3") lets you store four values for a total of 12 readings. Unless you REALLY need to, don't do it. The memory recall associated with each button does a rollover—goes back to the first recorded temperature—after you press the button four times, and keeping track of what 12 temperatures mean—yikes. Since you only recorded two temperatures on button 1, then if you press that button two more times, the first temperature (onset melting) should appear. Press it again, and the second temperature (end of melting) should appear. These would be your two temperature readings, and a difference of 2°C or less indicates a pure compound (Fig. 12.6).

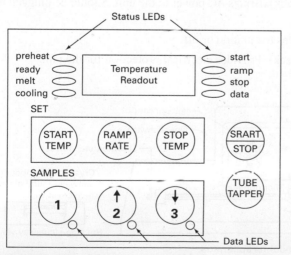

FIGURE 12.6 DigiMelt control panel.

If somebody else has used the unit before you, before you follow the instructions above or on the front of the unit:

1. Turn the unit off, wait about 2 seconds and turn it on again. This is the only way to clear out the data from the last run. (The "data" LED goes out.) Sure, if you enter your data it'll overwrite his, but with data from another run stored in the memory, it's too easy to confuse both sets. I'd clear the data if I were you.

2. If left alone, the unit will cool to room temperature, and NOT to the START TEMP that was programmed in the previous run. Why is this important? Let's say your entire class has recrystallized aspirin, M.P. 136°C, and the person before you chose 120°C as a START TEMP, a RAMP RATE of 5°C/min, and a STOP TEMP of 145°C, and just left after removing the sample and recording the temperature data. You turn the unit off-then-on to clear out that data, press START TEMP, RAMP RATE, and STOP TEMP in turn and decide that those values are acceptable. The unit says it is "cooling" and will continue to cool down to room temperature and NOT stop at the START TEMP for a new sample for you. You'll have to wait until the block temperature does become less than the START TEMP (but not by much), and press the START/STOP button to get to the "preheat" mode, let the block come up to your START TEMP, and then press START/STOP again to do the melting point experiment.

THE FISHER-JOHNS APPARATUS

The Fisher-Johns apparatus (Fig. 12.7) is different in that you don't use capillary tubes to hold the sample. Instead, you sandwich your sample between two round microscope cover slips (thin windows of glass) on a heating block. This type of melting-point apparatus is called a **hot stage**. It comes complete with spotlight. Look for the following.

1. *Line cord* (at the back). Brings AC power to the unit. Should be plugged into a live wall socket.

2. *On–off switch.* Turns the unit on or off.

3. *Fuse* (also at the back). Provides electrical protection for the unit.

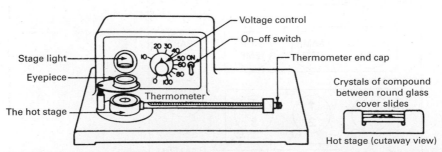

FIGURE 12.7 The Fisher-Johns apparatus.

4. *Voltage control.* Controls the *rate* of heating, *not the temperature*! The higher the setting, the faster the temperature rise.

5. *Stage light.* Provides illumination for samples.

6. *Eyepiece.* Magnifies the sample.

7. *Thermometer.* Gives the temperature of the sample.

8. *Thermometer end cap.* Keeps the thermometer from falling out. If the cap becomes loose, the thermometer tends to go belly-up and the markings turn over. Don't try to fix this while the unit is hot. Let it cool so you won't get burned.

9. *The hot stage.* This is the heating block on which samples are melted.

Operation of the Fisher-Johns Apparatus

1. *Don't assume that the unit is cold.* That is a good way to get burned. Read the temperature, and wait for the heating block to cool if the temperature is fewer than 20°C below the melting point of your compound. You'll have to wait until the block is at room temperature if you have no idea of the approximate melting point of your compound.

2. Keep your grubby fingers off the cover slips. Use tweezers or forceps.

3. Place a clean round glass cover slide in the well on the hot stage. *Never melt any samples directly on the metal stage.* Ever!

4. Put a few crystals on the glass. Not too many. As long as you can see them melt, you're all right.

5. Put another cover slide on top of the crystals to make a sandwich.

6. Set the voltage control to zero, if it's not already there.

7. Turn the on–off switch to ON. The light source should illuminate the sample. If not, call for help!

8. Now science turns into art. Set the *voltage control to any* convenient setting. The point is to get up to *within 20°C* of the *supposed* melting point. Yep, that's right. If you have no idea of the melting point, it may require several runs as you keep skipping past the point with a temperature rise of 5–10°C per minute. A convenient setting is *40*. This is just a suggestion, not an article of faith.

9. After you've melted a sample, let it cool, and remove the sandwich of sample and cover slides. *Throw it away!* Use an appropriate waste container.

10. Once you have an idea of the melting point (or have looked it up, or you were told), *get a fresh sample,* and bring the temperature up quickly at about *5–10°C per minute to within 20°C* of this approximate melting point. Then turn down the *voltage control* to get a *2°C-per-minute* rise. Patience!

11. When the first crystals *just start to melt,* record the temperature. When the last crystal *just disappears,* record the temperature. If both points appear to be the same, either the sample is extremely pure or the temperature rise was *too fast.*

12. Turn the on–off switch to OFF. Now set the voltage control to zero.

13. Let the stage cool, then remove the sandwich.

THE THOMAS-HOOVER APPARATUS

The Thomas-Hoover apparatus (Fig. 12.8) is the electromechanical equivalent of the Thiele tube or open beaker and hot oil methods (see "Using the Thiele Tube" later in this chapter). It has lots of features, and you should look for the following.

1. *Light box.* At the top of the device toward the back, a box holds a fluorescent lightbulb behind the thermometer. On the right side of this box are the fluorescent light switches.

    I'm told that there are versions of this beast that don't have a light box. I could write another whole set of instructions for that type, or, if you have such a unit, you could try ignoring anything that pertains to the light box.

2. *Fluorescent light switches.* Two buttons. Press and hold the red button down for a bit to light the lamp; press the black button to turn the lamp off.

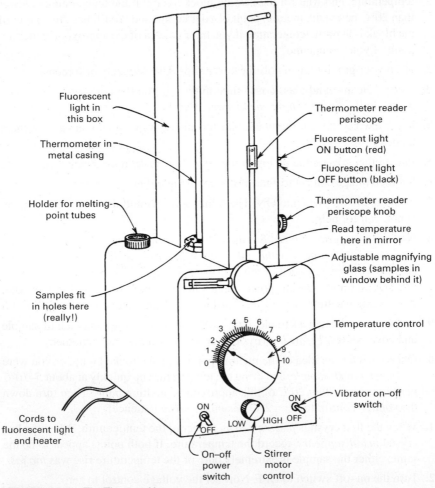

FIGURE 12.8 The Thomas-Hoover apparatus.

3. ***Thermometer.*** A special 300°C thermometer in a metal jacket is immersed in the oil bath that's in the lower part of the apparatus. Two slots have been cut in the jacket to let light illuminate the thermometer scale from behind and to let a thermometer periscope read the thermometer scale from the front.

4. ***Thermometer periscope.*** In front of the thermometer, this periscope lets you read a small magnified section of the thermometer scale. By turning the small knob at the lower right of this assembly, you track the movement of the mercury thread, and an image of the thread and temperature scale appears in a stationary mirror just above the sample viewing area.

5. ***Sample viewing area.*** A circular opening is cut in the front of the metal case so that you can see your samples in their capillary tubes (and the thermometer bulb), all bathed in the oil bath. You put the tubes into the oil bath through the holes in the capillary tube stage.

6. ***Capillary tube stage.*** In a semicircle around the bottom of the jacketed thermometer, yet behind the thermometer periscope, are five holes through which you can put your melting-point capillaries.

7. ***Heat.*** Controls the rate of heating, not the temperature. The higher the setting, the faster the temperature rise. At Hudson Valley Community College we've had a stop put in, and you can turn the dial only as far as the number 7. When it gets up to 10, you always smoke the oil. Don't do that.

8. ***Power on–off switch.*** Turns the unit on or off.

9. ***Stirrer motor control.*** Sets the speed of the stirrer from low to high.

10. ***Vibrator on–off switch.*** Turns the vibrator on or off. It's a spring-return switch, so you must hold the switch in the ON position. Let go, and it snaps to OFF.

11. ***Line cords.*** One cord brings AC power to the heater, stirrer, sample light, and vibrator. The other cord brings power to the fluorescent light behind the thermometer. Be sure that both cords are plugged into live wall sockets.

Operation of the Thomas-Hoover Apparatus

1. If the fluorescent light for the thermometer is not lit, press the red button at the right side of the light box, and hold it down for a bit to start the lamp. The lamp should remain lit after you release the button.

2. Look in the thermometer periscope, turn the small knob at the lower right of the periscope base, and adjust the periscope to find the top of the mercury thread in the thermometer. Read the temperature. Wait for the oil bath to cool if the temperature is fewer than 20°C below the approximate melting point of your compound. You'll have to wait until the bath is at room temperature if you have no idea of the melting point. You don't want to plunge your sample into oil that is so hot it might melt too quickly or melt at an incorrect temperature.

3. Turn the voltage control to zero, if it isn't there already.

4. Turn the power on–off switch to ON. The oil bath should become illuminated.

5. Insert your capillary tube in one of the capillary tube openings in the capillary tube stage. *This is not simple.* Be careful. If you snap a tube at this point, the entire unit may have to be taken apart to remove the pieces. You have to angle the tube toward the center opening and angle the tube toward you (as you face the instrument) at the same time (Fig. 12.9). The capillary tube has to negotiate a second set of holes, so you have to angle the tube toward the center.

6. Adjust the magnifying glass for the best view of your sample.

7. Turn the stirrer knob so that the mark on the knob is about halfway between the slow and fast markings on the front panel. That's just a suggestion. I don't have any compelling reasons for it.

8. Adjust the thermometer periscope to give you a good view of the top of the mercury thread in the thermometer.

9. Now science turns into art. Set the *heat control* to *any* convenient setting. The point is to get up to *within 20°C* of the *supposed* melting point. If you have no idea of the melting point, it may require several runs as you keep skipping past the point with a temperature rise of 5–10°C per minute. A convenient setting is *4.* This is just a suggestion, not an article of faith.

10. Remember, you'll have to keep adjusting the thermometer periscope to keep the top of the mercury thread centered in the image.

11. After you've melted a sample, *throw it away*!

12. Once you have an idea of the melting point (or looked it up, or were told), *get a fresh sample* and bring the temperature up quickly at about *5–10°C per minute to within 20°C* of this approximate melting point. Then turn down the *heat control* to get a *2°C-per-minute rise.* Patience!

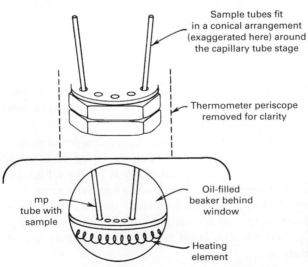

Sample tubes fit in a conical arrangement (exaggerated here) around the capillary tube stage

Thermometer periscope removed for clarity

mp tube with sample

Oil-filled beaker behind window

Heating element

FIGURE 12.9 Close-up of the viewing system for the Thomas-Hoover apparatus.

13. When the first crystals *just start to melt,* record the temperature. When the *last crystal just disappears,* record the temperature. If both points appear to be the same, either the sample is extremely pure or the temperature rise was *too fast.* If you record the temperature with the horizontal index line in the mirror matched to the lines etched on both sides of the periscope window and the top of the mercury thread at the same time, you'll be looking at the thermometer scale head-on. This will give you the smallest error in *reading* the temperature (Fig. 12.10).

14. Don't turn the control much past 7. You can get a bit beyond 250°C at that setting, and that should be *plenty* for any solid compound that you might prepare in organic lab. Above this setting, there's a real danger of smoking the oil.

15. Turn the power switch to OFF. You can also set the *heat control* to zero for the next person.

16. Press the black button on the right side of the light box, and turn the fluorescent light off.

17. Remove all capillary tubes and dispose of them properly.

There are a few more electric melting-point apparatuses around, and most of them work in the same way. A **sample holder, magnifying eyepiece**, and **voltage control** are common; an apparently essential feature of these devices is that dial markings are almost *never* temperature settings. That is, a setting of **60** will not give a temperature of 60°C but probably one that is much higher.

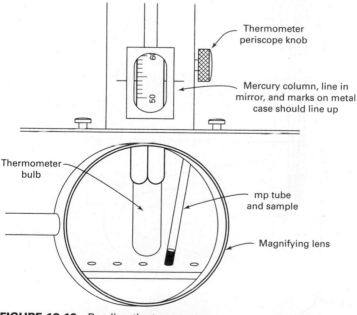

FIGURE 12.10 Reading the temperature.

USING THE THIELE TUBE

With the Thiele tube (Fig. 12.11), you use hot oil to transfer heat evenly to your sample in a melting-point capillary, just like the metal block of the Mel-Temp apparatus does. You heat the oil in the side arm and it expands. The hot oil goes up the side arm, warming your sample and thermometer as it touches them. Now the oil is cooler, and it falls to the bottom of the tube, where it is heated again by a burner. This cycle goes on automatically as you do the melting-point experiment in the Thiele tube.

Don't get any water in the tube, or when you heat the tube the water can boil and throw hot oil out at you. Let's start from the beginning.

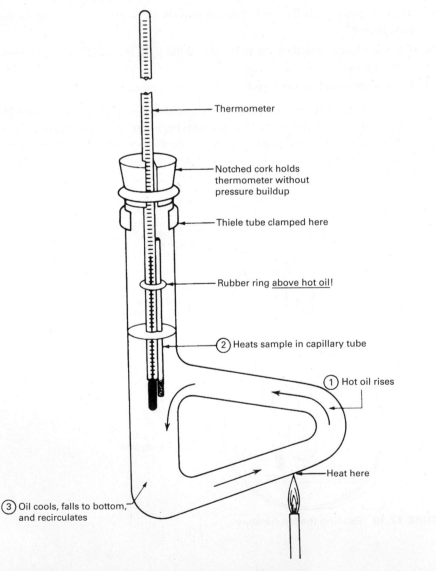

Thermometer

Notched cork holds thermometer without pressure buildup

Thiele tube clamped here

Rubber ring <u>above hot oil</u>!

(2) Heats sample in capillary tube

(1) Hot oil rises

Heat here

(3) Oil cools, falls to bottom, and recirculates

FIGURE 12.11 Taking melting points with the Thiele tube.

Cleaning the Tube

This is a bit tricky, so don't do it unless your instructor says so. Also, check with your instructor *before* you put fresh oil in the tube.

1. Pour the old oil out into an appropriate container, and let the tube drain.

2. Use a hydrocarbon solvent (hexane, ligroin, petroleum ether—and *no flames!*) to dissolve the oil that's left.

3. Get out the old soap and water and elbow grease, clean the tube, and rinse it out really well.

4. Dry the tube thoroughly in a drying oven (usually >100°C). Carefully take it out of the oven, and let it cool.

5. Let your instructor examine the tube. If you get the OK, *then* add some fresh oil. Watch it. First, *no water.* Second, don't overfill the tube. Normally, the oil expands as you heat the tube. If you've overfilled the tube, oil will crawl out and get you.

Getting the Sample Ready

Here you use a loaded melting-point capillary tube (see "Loading the Melting-Point Tube" earlier in the chapter) and attach it directly to the thermometer. Unfortunately, the thermometer has bulges; there are some problems, and you may snap the tube while attaching it to the thermometer.

1. Get a thin rubber ring or cut one from a piece of rubber tubing.

2. Put the *bottom* of the loaded mp tube *just above* the place where the thermometer constricts (Fig. 12.12), and carefully roll the rubber ring onto the mp tube.

3. Reposition the tube so that the sample is near the center of the bulb and the rubber ring is near the open end. *Make sure the tube is vertical.*

4. Roll the rubber ring up to near the top of the mp tube, out of harm's way.

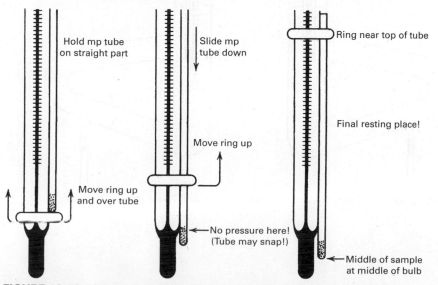

FIGURE 12.12 Attaching the mp tube to the thermometer without a disaster.

Dunking the Melting-Point Tube

There are more ways of keeping the thermometer suspended in the oil than I care to list. You can cut or file a notch on the side of the cork, drill a hole, and insert the thermometer (*be careful!*). Finally, cap the Thiele tube and dunk the mp tube (Fig. 12.11). The notch is there so that pressure will not build up as the tube is heated. *Keep the notch open, or the setup may explode.*

But this requires drilling or boring corks, something you try to avoid. (Why have ground-glass jointware in the undergraduate lab?) You can *gently* hold a thermometer and a cork in a clamp (Fig. 12.13). Not too much pressure, though!

Finally, you might put the thermometer in the **thermometer adapter** and suspend that, clamped gently by the rubber part of the adapter, not by the ground-glass end. Clamping ground glass will score the joint.

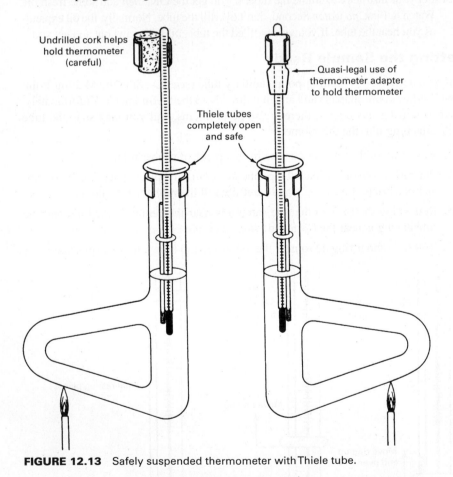

Undrilled cork helps hold thermometer (careful)

Quasi-legal use of thermometer adapter to hold thermometer

Thiele tubes completely open and safe

FIGURE 12.13 Safely suspended thermometer with Thiele tube.

Heating the Sample

The appropriately clamped thermometer is set up in the Thiele tube as in Fig. 12.11. Look at this figure *now,* and remember to heat the tube *carefully—always carefully—at the elbow.* Then,

1. If you don't know the melting point of the sample, heat the oil fairly quickly, *but no more than 10°C per minute,* to get a rough melting point. And it will be rough indeed, since the temperature of the thermometer usually lags that of the sample.

2. After this sample has melted, lift the thermometer and attached sample tube *carefully (it may be HOT)* by the thermometer up at the clamp, until they are *just out of the oil.* This way the thermometer and sample can cool, and the hot oil can drain off. Wait for the thermometer to cool to about room temperature before you remove it entirely from the tube. Wipe off some of the oil, reload a melting-point tube (*never remelt melted samples*), and try again. And heat at 2°C per minute this time.

EXERCISES

1. "If you can see it, you can see it melt." This should put the correct amount of a solid in a melting-point tube. If someone has put about one-quarter of an inch of solid in the tube, should they go for an eye exam?

2. Define "melting point."

3. How fast should a melting-point capillary sample be heated?

4. Why not remelt samples? Don't they just cool and solidify in exactly the reverse manner that they heated and melted?

5. Define "freezing point."

Chapter 13

RECRYSTALLIZATION

■ *Too much solvent and your crystals don't come back.*

The essence of a recrystallization is **purification**. Messy, dirty compounds are cleaned up, purified, and can then hold their heads up in public again. The sequence of events you use will depend a lot on how messy your crude product is and on just how soluble it is in various solvents.

In any case, you'll have to remember a few things.

1. Find a solvent that will *dissolve the solid while hot.*

2. The same solvent *should not dissolve it while cold.*

3. The *cold solvent* must keep impurities dissolved in it *forever or longer.*

This is the major problem. And it requires some experimentation. That's right! Once again, art over science. Usually, you'll know what you should have prepared, so the task is easier. It requires a trip to **your notebook** and, possibly, the **internet** (see Chapter 2, "Keeping a Notebook," and Chapter 3, "Mining Your Own Business"). You have the data on the solubility of the compound in your notebook. What's that, you say? *You don't have the data in your notebook?* Congratulations, you get the highest F in the course.

Information in the notebook (which came from the internet) for your compound might say, for alcohol (meaning *ethyl* alcohol), **s.h.** Since this means **s**oluble in **h**ot alcohol, it implies that it is **i**nsoluble (**i.**) in cold alcohol. Then alcohol is probably a good solvent for recrystallization of that compound. Also, check on the **color** or **crystalline form**. This is important since:

1. A color in a supposedly white product is an impurity.
2. A color in a colored product is *not* an impurity.
3. The *wrong color* in a product is an impurity.

You can usually assume that impurities are present in small amounts. Then you don't have to guess what possible impurities might be present or what they might be soluble or insoluble in. If your sample is really dirty, the assumption can be fatal. This doesn't usually happen in an undergraduate lab, but you should be aware of it.

FINDING A GOOD SOLVENT

If the solubility data for your compound is not readily available, then:

1. Place 0.10 g of your solid (weighed to 0.01 g) in a test tube.
2. Add 3 mL of a solvent, stopper the tube, and shake the bejesus out of it. If *all of the solid dissolves at room temperature,* then your solid is **soluble**. Do *not* use this solvent as a recrystallization solvent. (You must make note of this in your notebook, though.)
3. If none (or very little) of the solid dissolved at room temperature, unstopper the tube and heat it (*careful—no flames and get a boiling stone!*) and shake it and heat it and shake it. You may have to heat the solvent to a gentle boil. (*Careful!* Solvents with low boiling points often boil away.) If it does *not* dissolve at all, then do not use this as a recrystallization solvent.
4. If the sample *dissolved when hot,* and *did not dissolve at room temperature,* you're on the trail of a good recrystallization solvent. One last test.
5. Place the test tube (no longer hot) in an ice-water bath, and cool it about 5°C or so. If lots of crystals come out, this is good, and this is your recrystallization solvent.
6. Suppose your crystals don't come back when you cool the solution. Get a glass rod into the test tube, stir the solution, rub the inside of the tube with the glass rod, and agitate that solution. If crystals still don't come back, perhaps you'd better find another solvent.
7. Suppose, after all this, you still haven't found a solvent. Look again. Perhaps your compound *completely* dissolved in ethanol at room temperature and would *not* dissolve in water. Aha! Ethanol and water are **miscible** (i.e., they mix in all proportions). You will have to perform a **mixed-solvent recrystallization** (see "Working with a Mixed-Solvent System" later in this chapter).

GENERAL GUIDELINES
FOR A RECRYSTALLIZATION

Here are some general rules to follow for purifying any solid compound.

1. Put the solid in an Erlenmeyer flask, not a beaker. If you recrystallize compounds in beakers, you may find the solid climbing the walls of the beaker to get at you as a reminder. A 125-mL Erlenmeyer usually works. Your solid should look comfortable in it, neither cramped nor with too much space. You probably shouldn't fill the flask more than one-fifth to one-fourth full.

2. Heat a large quantity of a proven solvent (see preceding section) to the boiling point, and *slowly add the hot solvent* to the sample in the Erlenmeyer. *Slowly!* A word about solvents: *Fire!* Solvents burn! *No flames!* A hot plate here would be better. You can even heat solvents in a *steam or water bath.* But—*no flames!*

3. Carefully add the hot solvent to the solid to just dissolve it. In parts. Your first shot of solvent *must* not dissolve all of the solid. Or else you might have gone too far and have more solvent in that flask than to just dissolve your solid. This can be tricky, since hot solvents evaporate, cool down, and so on. Ask your instructor.

4. Add a slight excess of the hot solvent to keep the solid dissolved.

5. If the solution is only slightly colored, the impurities will stay in solution. Otherwise, the big gun, **activated charcoal**, may be needed (see "Activated Charcoal" later in this chapter). Remember, if you were working with a colored compound, it would be silly to try to get rid of all the color, since you would get rid of all the compound and probably all your grade.

6. Keep the solvent hot (*not boiling*) and look carefully to see if there is any trash in the sample. This could be old boiling stones, sand, floor sweepings, and so on. Nothing you'd want to bring home to meet the folks. *Don't confuse real trash with undissolved good product!* If you add more hot solvent, good product will dissolve, and trash will not. If you have trash in the sample, do a **gravity filtration** (see following section).

7. Let the Erlenmeyer flask and the hot solution cool. Slow cooling gives better crystals. Garbage doesn't get trapped in them. But this can take what seems to be an interminable length of time. (I know, the entire lab seems to take an interminable length of time.) So, after the flask cools and it's just *warm* to the touch, then put the flask in an ice-water bath to cool. *Watch it!* The flasks have a habit of turning over in the water baths and letting all sorts of water destroy all your hard work! Also, a really hot flask will shatter if plunged into the ice bath, so again, watch it.

Why everyone insists on loading up a bucket with ice and trying to force a flask into this mess, I'll never know. How much cooling do you think you're going to get with just a few small areas of the flask barely touching ice?

Ice bath really means ice-water bath.

8. **"Is it cool enough yet?"** Hey, cool it, yourself. When the solution gets down to the temperature of the ice-water bath (and you should know what that is, eh?), then filter the crystals on a **Buchner funnel** or **Hirsch funnel**.

9. Dry them and take a melting point, as described in Chapter 12, "The Melting-Point Experiment."

My Product Disappeared

If you don't get any crystals back after cooling, remember: They're in there. Now is NOT the time to despair and throw anything away. First, see about getting a glass rod down into the flask, and, with a bit of vigorous scratching of the flask walls and stirring of the solution, crystals may appear. Waiting awhile after the stirring may help, but this is problematic; you can't wait all day.

You could just have too much solvent, and the crystals won't come out. Time to evaporate some of the solvent—start with about half—then go back to #7 above, cooling the solution again. Still no crystals? Don't throw anything away. You might have to evaporate solvent again. Remember—they're in there . . .

GRAVITY FILTRATION

If you find yourself with a flask full of hot solvent and your product dissolved in it, along with all sorts of trash, this is for you. You'll need more hot solvent, a ring stand with a ring attached, possibly a clay triangle, some filter paper, a clean, dry flask, and a **stemless funnel**. Here's how **gravity filtration** works.

1. Fold up a **filter cone** from a piece of filter paper (Fig. 13.1). It should fit nicely, within a single centimeter or so of the top of the funnel. For those who wish to filter with more panache, try using **fluted filter paper** (see "Finishing the final fluted fan," Fig. 13.12).

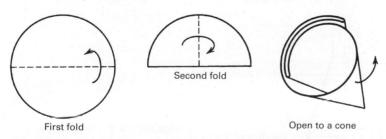

First fold   Second fold   Open to a cone

FIGURE 13.1   Folding filter paper for gravity filtration.

2. Get yourself a **stemless funnel** or, at least, a **short-stem funnel**, or even a **powder funnel**. Why? Go ahead and *use* a long-stem funnel and watch the crystals come out in the stem as the solution cools, blocking up the funnel (Fig. 13.2).

3. Put the **filter paper cone** in the **stemless funnel**.

4. Support this in a ring attached to a ring stand (Fig. 13.3). If the funnel is too small and you think it could fall through the ring, you may be able to get a **wire** or **clay triangle** to support the funnel in the ring (Fig. 13.4).

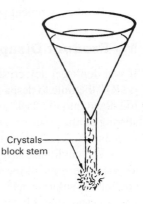

Crystals block stem

FIGURE 13.2 The "too long a funnel stem—oops!" problem.

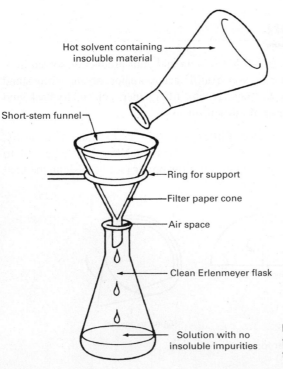

Hot solvent containing insoluble material

Short-stem funnel

Ring for support

Filter paper cone

Air space

Clean Erlenmeyer flask

Solution with no insoluble impurities

FIGURE 13.3 The gravity filtration setup with a funnel that fits the iron ring.

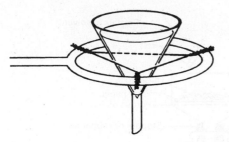

FIGURE 13.4 A wire triangle holding a small funnel in a large iron ring for gravity filtration.

5. Put the new, *clean, dry flask* under the funnel to catch the hot solution as it comes through. All set?

6. Get that flask with the solvent, product, and trash hot again. (*No flames!*) You should get some fresh, clean solvent hot as well. (*No flames!*)

7. Carefully pour the hot solution into the funnel. As it is, some solvents evaporate so quickly that product will probably come out on the filter paper. It is often hard to tell the product from the insoluble trash. Then—

8. Wash the filter paper down with *a little hot solvent*. The product will redissolve. The trash won't.

9. Now, you let the *trash-free* solution cool, and clean crystals should come out. Since you have probably added solvent to the solution, *don't be surprised if no crystals come out of solution. Don't panic, either!* Just boil away some of the solvent (not out into the room air, please), let your solution cool, and wait for the crystals again. If they *still* don't come back, just repeat the boiling.

Do not boil to dryness!

Somehow, lots of folks think recrystallization means dissolving the solid, then boiling away all the solvent to dryness. No! There must be a way to convince these lost souls that *the impurities will deposit on the crystals.* After the solution has cooled, crystals come out, sit on the bottom of the flask, and *must be covered by solvent!* Enough solvent to keep those nasty impurities dissolved and off the crystals.

THE BUCHNER FUNNEL
AND FILTER FLASK

The **Buchner funnel** (Fig. 13.5) is used primarily for separating crystals of product from the liquid above them. If you have been *boiling your recrystallization solvents dry, you should be horsewhipped* and forced to reread these sections on recrystallization!

1. Get a piece of filter paper large enough to cover all of the holes in the bottom plate, and yet *not* curl up the sides of the funnel. It is placed *flat* on the plate (Fig. 13.5).

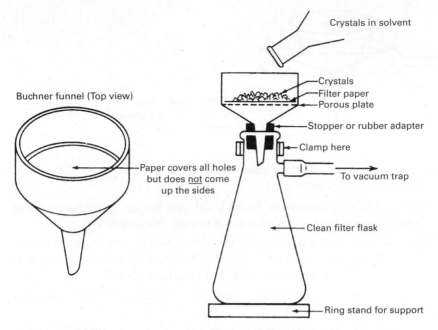

FIGURE 13.5 The Buchner funnel at home and at work.

2. Clamp a **filter flask** to a ring stand. This filter flask, often called a **suction flask**, is a very heavy-walled flask that has a side arm on the neck. A piece of heavy-walled tubing connects this flask to the **water trap** (see Fig. 13.6).

3. Now use a **rubber stopper** or **filter adapter** to stick the Buchner funnel into the top of the filter flask. The Buchner funnel makes the setup top-heavy and prone to be prone—and broken. Clamp the flask first, or go get a new Buchner funnel to replace the one you'll otherwise break.

4. The **water trap** is in turn connected to a source of vacuum, most likely a **water aspirator** (see Fig. 13.7).

5. The faucet on the **water aspirator** should be turned on *full blast!* This should suck down the filter paper, which you now *wet with some of the* **cold** *recrystallization solvent.* This will make the paper stick to the plate. You may have to push down on the Buchner funnel a bit to get a good seal between the rubber adapter and the funnel.

6. Swirl and pour the crystals and solvent *slowly and directly into the center of the filter paper,* as if to build a small mound of product there. *Slowly!* Don't flood the funnel by filling it right to the brim and waiting for the level to go down. If you do that, the paper may float up, ruining the whole setup.

7. Use a very small amount of the same *cold* recrystallization solvent and a spatula to remove any crystals left in the flask. Then you can use some of the *fresh,*

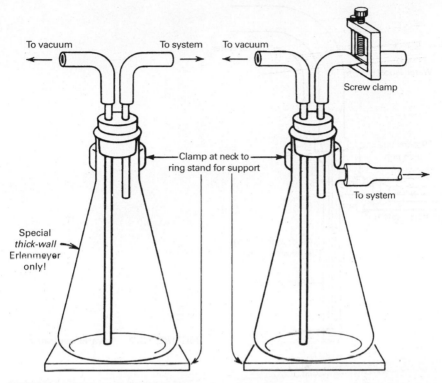

FIGURE 13.6   A couple of water traps hanging around.

cold recrystallization solvent and slowly pour it over the crystals to wash away any old recrystallization solvent and dissolved impurities.

8. Leave the aspirator on, and let air pass through the crystals to help them dry. You can put a thin rubber sheet, a **rubber dam**, over the funnel. The vacuum pulls it in, and the crystals are pressed clean and dry. You won't have air or moisture blowing through, and possibly decomposing, your product. Rubber dams are neat.

9. When the crystals are dry *and you have a* **water trap**, just turn off the water aspirator. Water won't back up into your flask. (If you've been foolhardy and have filtered without a water trap, just remove the rubber tube connected to the filter flask side arm [Fig. 13.5].)

10. At this point, you may have a *cake of crystals* in your **Buchner funnel**. The easiest way to handle this is to *carefully lift the cake* of crystals out of the funnel *along with the filter paper,* plop the whole thing onto a larger piece of filter paper, and let everything dry overnight. If you are pressed for time, *scrape the damp filter cake from the filter paper, but don't scrape any filter paper fibers into the crystals.* Repeatedly press the crystals between dry sheets of

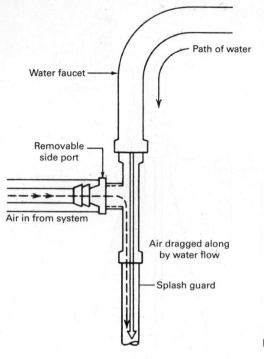

Path of water

Water faucet

Removable side port

Air in from system

Air dragged along by water flow

Splash guard

FIGURE 13.7 A water aspirator.

filter paper, and change sheets until the crystals no longer show any solvent spot after pressing. Those of you who use **heat lamps** may find your white crystalline product turning into instant charred remains.

11. When your cake is *completely dry,* weigh a vial, put in the product, and weigh the vial again. Subtracting the weight of the vial from the weight of the vial and sample gives the weight of the product. This **weighing by difference** is easier and less messy than weighing the crystals directly on the balance. This weight should be included in the label on your **product vial** (see Chapter 5, "Microscale Jointware," and Chapter 11, "On Products").

Just a Note

I've said that a Buchner funnel is used primarily for separating crystals of product from the liquid above them. And in the section on drying agents, I tell people to use a gravity filtration setup to separate a drying agent from a liquid product. Recently, I've had some people get the notion that you can Buchner-filter liquid products from drying agents. I don't advise that. You will probably lose a lot of your product, especially if it has a low boiling point (<100°C). Under this *vacuum filtration,* your product simply evaporates along with your grade.

THE HIRSCH FUNNEL AND FRIENDS

You can think of the **Hirsch funnel** (Fig. 13.8) as a smaller, sloping-sided Buchner funnel, and, for the most part, you use it in the same way.

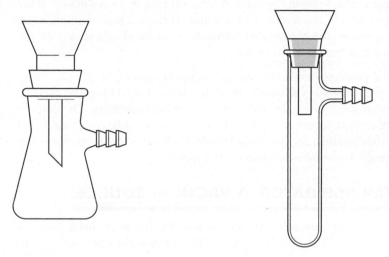

FIGURE 13.8 The Hirsch funnel and friends.

ACTIVATED CHARCOAL

Activated charcoal is ultrafinely divided carbon with lots of places to suck up big, huge, polar, colored impurity molecules. Unfortunately, if you use too much, it'll suck up *your product!* And, if your product was white or yellow, it'll have a funny gray color from the excess charcoal. Sometimes the impurities are untouched and only the product is absorbed. Again, it's a matter of trial and error. Try not to use too much. Suppose you've got a *hot solution* of some solid, *and the solution is highly colored.* Well,

1. First, *make sure your product should not* **be** *colored!*
2. Take the flask with your filthy product off the heat, and swirl the flask. This dissipates any superheated areas so that when you add the activated charcoal, the solution doesn't foam out of the flask and onto your shoes.
3. *Add the activated charcoal.* Put a small amount, about the size of a pea, on your spatula; then throw the charcoal in. Stir. The solution should turn black. Stir and heat.
4. Set up the **gravity filtration** and filter off the carbon. It is especially important to *wash off any product caught on the charcoal,* and it is really hard to see anything here. You should take advantage of **fluted filter paper**. It should give a more efficient filtration.

5. Yes, have some extra fresh solvent heated as well. You'll need to add a few milliliters of this to the hot solution to help keep the crystals from coming out on the filter paper. And you'll need more to help wash the crystals off the paper when they come out on it anyway.

6. The filtered solution should be *much cleaner* than the original solution. If not, you'll have to *add charcoal and filter again.* There is a point of diminishing returns, however, and one or two treatments is usually all you should do. Get some guidance from your instructor.

Your solid products should not be gray. Liquid products (yes, you can do liquids!) will let you know that you didn't get all the charcoal out. Often, you can't see charcoal contamination in liquids while you're working with them. The particles stay suspended for awhile, but after a few days, you can see a layer of charcoal on the bottom of the container. Sneaky, those liquids. By the time the instructor gets to grade all the products—*voilà*—the charcoal has appeared.

THE WATER ASPIRATOR: A VACUUM SOURCE

Sometimes you'll need a vacuum for special work like **vacuum distillation** and **vacuum filtration** as with the Buchner funnel. The **water aspirator** is an inexpensive source of vacuum (Fig. 13.7).

When you turn the water on, the water flow draws air in from the side port on the aspirator. The faster the water goes through, the faster the air is drawn in. Pretty neat, huh? I've shown a plastic aspirator, but many of the older metal varieties are still around.

You may have to pretest some aspirators before you find one that will work well. It'll depend on the water pressure in the pipes, too. Even the number of people using aspirators on the same water line can affect the performance of these devices. You can test them by going to an aspirator and turning the faucet on *full blast.* It does help to have a sink under the aspirator. If water leaks out the side port, *tell your instructor and find another aspirator.* Wet your finger and place it over the hole in the side port to feel if there is any vacuum. If there is *no vacuum,* tell your instructor and find another aspirator. Some of these old, wheezing aspirators produce a very weak vacuum. You must decide for yourself if the suction is "strong enough." There should be a **splash guard** or rubber tubing leading the water stream directly into the sink. This will keep the water from going all over the room. If you check and don't find such protection, see your instructor. All you have to do with a fully tested and satisfactory aspirator is to hook it up to the **water trap**.

THE WATER TRAP

Every year, I run a chem lab, and when someone is doing a **vacuum filtration**, suddenly I'll hear a scream and a moan of anguish, as water backs up into someone's filtration system. Usually there's not much damage, since the filtrate in the suction flask is generally thrown out. For **vacuum distillations**, however, this **suck-back** is disaster. It happens whenever there's a pressure drop on the water line big enough to cause the flow to decrease so that there is a *greater vacuum in the system than in the aspirator.* Water, being water, flows into the system. Disaster.

So, for your own protection, make up a **water trap** from some stoppers, rubber tubing, a thick-walled Erlenmeyer or filter flask, and a screw clamp (Fig. 13.6). *Do not use garden-variety Erlenmeyers; they may implode without warning.* Two versions are shown. I think the setup using the filter flask is more flexible. The screw clamp allows you to let air into your setup at a controlled rate. You might clamp the water trap to a ring stand when you use it. The connecting hoses have been known to flip unsecured flasks two out of three times.

WORKING WITH A MIXED-SOLVENT SYSTEM—THE GOOD PART

If, after sufficient agony, you cannot find a single solvent to recrystallize your product from, you may just give up and try a *mixed-solvent system.* Yes, it does mean you mix more than one solvent and *recrystallize using the mixture.* It should only be so easy. Sometimes you are told what the mixture is and the correct proportions. Then it is easy.

For an example, I could use "solvent 1" and "solvent 2," but that's clumsy. So I'll use the ethanol–water system and point out the interesting stuff as I go along.

The Ethanol—Water System

If you look up the solubility of water in ethanol (or ethanol in water), you find an ∞. This means they mix in all proportions. Any amount of one dissolves completely in the other—no matter what. Any volumes, any weights. You name it. The special word for this property is **miscibility**. Miscible solvent systems are the kinds you should use for mixed solvents. They keep you out of trouble. You'll be adding amounts of water to the ethanol, and ethanol to the water. If the two weren't miscible, they might begin to separate and form two layers as you changed the proportions. Then you'd have REAL trouble. So, go ahead. You *can* work with mixtures of solvents that aren't miscible. But don't say you haven't been warned.

The ethanol–water mixture is useful because:

1. *At high temperatures, it behaves like alcohol!*
2. *At low temperatures, it behaves like water!*

From this, you should get the idea that it would be good to use a mixed solvent to recrystallize compounds that are *soluble in alcohol* yet *insoluble in water.* You see, each solvent alone cannot be used. If the material is soluble in alcohol, not many crystals come back from alcohol alone. If the material is insoluble in water, you cannot even begin to dissolve it. So, you have a *mixed solvent,* with the best properties of *both* solvents. To actually perform a *mixed-solvent recrystallization,* you:

1. Dissolve the compound in the smallest amount of *hot ethanol.*
2. Add *hot water* until the solution turns cloudy. This **cloudiness** is *tiny crystals of compound coming out of solution.* Heat this solution to dissolve the crystals. If they do not dissolve completely, add *a very little hot ethanol* to force them back into solution.
3. Cool and collect the crystals on a **Buchner funnel**.

Any solvent pair that behaves the same way can be used. The addition of hot solvents to one another can be tricky. It can be *extremely dangerous* if the boiling points of the solvents are very different. For the *water–methanol mixed solvent,* if 95°C water hits *hot methanol* (bp 65.0°C), watch out!

There are other miscible, mixed-solvent pairs—petroleum ether and diethyl ether, methanol and water, and ligroin and diethyl ether among them.

A MIXED-SOLVENT
SYSTEM—THE BAD PART

Every silver lining has a cloud. More often than not, compounds "recrystallized" from a mixed-solvent system don't form crystals. Your compound may form an *oil* instead.

Oiling out is what it's called; more work is what it means. Compounds usually oil out if *the boiling point of the recrystallization solvent is higher than the melting point of the compound,* although that's not the only time. In any case, if the oil solidifies, the impurities are trapped in the now-solid "oil," and you'll have to purify the solid again.

Don't think you won't ever get oiling out if you stick to single, unmixed solvents. It's just that with two solvents, there's a greater chance you'll hit on a composition that will cause this.

Temporarily, you can:

1. Add more solvent. If it's a mixed-solvent system, try adding more of the solvent the solid is NOT soluble in. Or add more of the OTHER solvent. No contradiction. The point is to *change the composition.* Whether single solvent or mixed solvent, changing the composition is one way out of this mess.

2. Redissolve the oil by heating; then shake up the solution as it cools and begins to oil out. When these smaller droplets finally freeze out, they may form crystals that are relatively pure. They may not. You'll probably have to clean them up again. Just don't use the same recrystallization solvent.

Sometimes, once a solid oils out it doesn't want to solidify at all, and you might not have all day. Try removing a sample of the oil with an eyedropper or a disposable pipet. Then get a glass surface (watch glass) and add a few drops of a solvent that the compound is known to be *insoluble* in (usually water). Then use the rounded end of a glass rod to *triturate the oil with the solvent.* **Trituration** can be described loosely as beating an oil into a crystalline solid. Then you can put these crystals back into the rest of the oil. Possibly they'll act as seed crystals and get the rest of the oil to solidify. Again, you'll still have to clean up your compound.

SALTING OUT

Sometimes you'll have to recrystallize your organic compound from water. No big deal. But sometimes your organic compound is more than ever-so-slightly insoluble in water, and not all the compound will come back. Solution? Salt solution! A pinch of salt in the water raises the **ionic strength**. There are now charged ions in the water. Some of the water that solvated your compound goes with the salt ions. Your organic compound does not particularly like charged ions anyway, so more of your organic compound comes out of the solution.

You can dissolve about 36 g of common salt in 100 mL of cold water. That's the upper limit for salt. You can estimate how much salt you'll need to practically saturate the water with salt. Be careful, though—if you use too much salt, you may find yourself collecting salt crystals along with your product. (See also the application of salting out when you have to do an extraction, "Extraction Hints," in Chapter 15.)

WORLD-FAMOUS FAN-FOLDED FLUTED PAPER

Some training in origami is *de rigueur* for chemists. It seems that the regular filter paper fold is inefficient, since very little of the paper is exposed. The idea here is to **flute** or **corrugate** the paper, increasing the surface area that is in contact with your filtrate. You'll have to do this several times to get good at it.

Right here let's review the difference between **fold** and **crease**. Folding is folding; creasing is folding, then stomping on it, and running fingers and fingernails over a fold over and over and over. Creasing so weakens the paper, especially near the folded point, that it may break at an inappropriate time in the filtration.

1. Fold the paper in half, then in half again, then in half again (Fig. 13.9). Press on this wedge of paper to get the fold lines to stay, but *don't crease. Do this in one direction only.* Always fold either toward you or away from you, but don't do both.

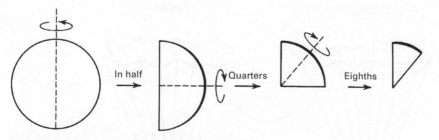

FIGURE 13.9 Folding filter paper into eighths.

2. Unfold this cone *twice* so it looks like a semicircle (Fig. 13.10), and then put it down on a flat surface. Look at it and think for not less than two full minutes the first time you do this.

3. OK. Now try a "fan fold." You alternately fold, first in one direction and then the other, every individual eighth section of the semicircle (Fig. 13.11).

4. Open the fan and play with it until you get a fairly fluted filter cone (Fig. 13.12).

5. It'll be a bit difficult, but try to find the two opposing sections that are not folded correctly. Fold them inward (Fig. 13.12), and you'll have a fantastic fan-folded fluted filter paper of your very own.

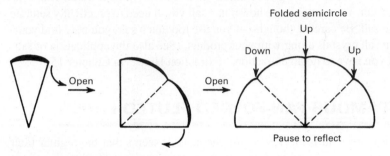

FIGURE 13.10 Unfolding to a sort of bent semicircle.

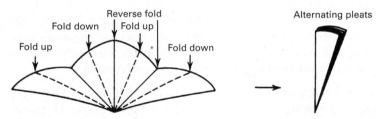

FIGURE 13.11 Refolding to a fan.

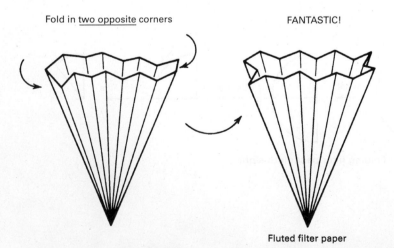

FIGURE 13.12 Finishing the final fluted fan.

P. S. For those with more money than patience, prefolded fan-folded fluted filter paper is available from suppliers.

EXERCISES

1. An impatient lab student plunges his hot, totally dissolved solution into his ice-water bath to get the crystals out quickly. Comment on the purity of his recrystallized product.

2. You perform the recrystallization procedure and get no crystals. What might you do to rectify this situation? Would it help to recall a Prego spaghetti sauce commercial that, talking about the composition of the sauce, stated, "It's in there!"?

3. OK, you have an oil, and you have to change the composition of your mixed-solvent system to get to a different point on the phase diagram. Do you add more of the more-polar solvent or the less-polar solvent?

Chapter 14

RECRYSTALLIZATION: Microscale

Microscale recrystallizations are done the same way the big ones are. You just use teeny-tiny glassware. Read the general section on crystallization first. Then come back here, and:

1. Put your solid in a small test tube.
2. Add a **microboiling stone** or a tiny **magnetic stirring vane** if you're on a stirring hot plate.
3. *Just cover* the solid with recrystallization solvent.
4. Heat this mix in the sand bath; stir with the spin bar or shake the test tube.
5. If the solid isn't dissolving, *consider*:
   a. The stuff not dissolving is insoluble trash; adding lots more solvent won't *ever* dissolve it.
   b. The stuff not dissolving is your compound; you need to add 1–2 more drops of solvent.
   This is a decision you'll have to make based on your own observations. Time to take a little responsibility for your actions. I wouldn't, however, add much more than 2–3 mL at the microscale level.
6. OK. Your product has *just dissolved* in the solvent. Add another 10% or so of excess solvent to keep the solid in solution during the hot filtration. This could be half a

drop or so. Careful. If you're going to use *decolorizing carbon,* add it now; the whole tube should turn black. Heat for 5–10 seconds, and then use Pasteur pipets to filter. You might have to filter twice if you've used carbon.

7. Get your solution into a test tube to cool and recrystallize. Let it cool slowly to room temperature; then put the tube in an ice-water bath for final recrystallization.

8. "I didn't get any crystals." Heat the tube and boil off some of the solvent. (*Caution—Bumping!*) Use a microboiling stone, and don't point the tube at anyone. Repeat steps 7 and 8 until you get crystals. The point is to get to *saturation* while hot (all the solid that can be dissolved in the solvent), and the crystals will come out when cold.

If you've compared this outline to the large-scale recrystallization, you'll find that, with one exception—keeping a reservoir of hot solvent ready—the only difference here is the size of the equipment. Microscale means test tubes for flasks, filter pipets for funnels with paper, and so on. Because you can easily remove solvent at this scale, adding too much solvent is not quite the time-consuming boiling-off process you'd have to do on a larger scale.

ISOLATING THE CRYSTALS

First, see the section on "Pipet Filtering—Solids" in Chapter 7 where you remove solvent with a pipet. I also mention using a Hirsch funnel. Just reread the section on the Buchner funnel filtration (Chapter 13, "Recrystallization"), and where you see "Buchner" substitute "Hirsch" and paste copies of the Hirsch funnel drawing over the Buchner funnel drawing. Just realize that the tiny disk of paper can fly up more easily. Think small.

CRAIG TUBE FILTRATION

Usually this is called Craig tube crystallization, because you've pipet-filtered your hot solution into the bottom of a Craig tube. So if you've recrystallized in something else, redissolve the crystals and get this solution into the bottom of a Craig tube (Fig. 14.1).

1. Carefully put the upper section of the Craig tube into the bottom section. Let the solution slowly cool to room temperature and then put the tube in an ice-water bath.

2. Now the fun part. You've got to turn this thing upside down, put it in a centrifuge tube, and *not* have it come apart and spill everything. Easy to say—"Solvent is now removed by inverting the Craig tube assembly into a centrifuge tube . . ." (Mayo *et al.*, p 80). Easy to do? Well . . .

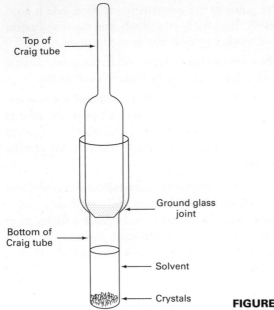

Top of Craig tube

Ground glass joint

Bottom of Craig tube

Solvent

Crystals

FIGURE 14.1 Crystals in Craig tube.

3. Get your centrifuge tube, and cut a length of copper wire as long as the tube. Make a loop in the end that's big enough to slip over the end of the Craig tube stem (Fig. 14.2). Twist the wire so that the loop won't open if you pull on it a bit.

4. Bend the loop so that it makes about a right angle from the wire (Fig. 14.2). Ready?

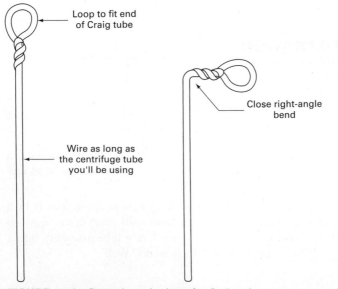

Loop to fit end of Craig tube

Close right-angle bend

Wire as long as the centrifuge tube you'll be using

FIGURE 14.2 Preparing wire loop for Craig tube.

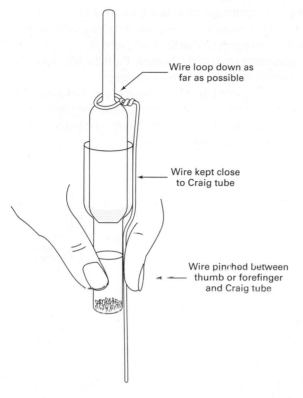

Wire loop down as
far as possible

Wire kept close
to Craig tube

Wire pinched between
thumb or forefinger
and Craig tube

FIGURE 14.3 Wiring the
Craig tube.

5. Hold the bottom of the Craig tube between your thumb and forefinger. Put the loop over the end of the tube (Fig. 14.3).

6. Hold the wire gently between your thumb (or forefinger) and the glass tube bottom. Grip the hanging wire, and gently pull it downward so that the Craig tube top is pulled into the bottom (Fig. 14.3). Too much force here and you could snap the stem. Watch it.

7. Once this is adjusted, tightly press the wire to the glass, so that the wire will keep the Craig tube closed when you turn the tube upside down.

8. Put the centrifuge tube upside down over the top of the Craig tube (Fig. 14.4*b*). Watch it. If the fit is snug, the copper wire can cause problems, such as breaking the tubes. If the centrifuge tube is too tight, you might consider getting a wider one.

9. Once you get to this point, you may have three situations:
    Baby Bear (Fig. 14.4*a*): The centrifuge tube is a bit shorter than the Craig tube. Some of the Craig tube peeks out from the centrifuge tube. The centrifuge *might* have clearance in the head to accommodate the extra length. I don't like this. Perhaps you should ask your instructor. Invert the setup and get ready to centrifuge.

Mama Bear (Fig. 14.4*b*): The centrifuge tube is a bit longer than the Craig tube. "A bit" here can vary. I mean you can push the Craig tube up into the centrifuge tube with your fingers so that the Craig tube stem touches the bottom or is within a centimeter or so of the bottom. Push the tube up as far as it will go. Invert the setup, and get ready to centrifuge.

Papa Bear (Fig. 14.4*c*): The centrifuge tube is a *lot* longer than the Craig tube. You can't push it up to the bottom with your fingers. Grab the copper wire at the end, invert the setup, and quickly *lower* the Craig tube to the bottom of the centrifuge tube. Now you're ready to centrifuge.

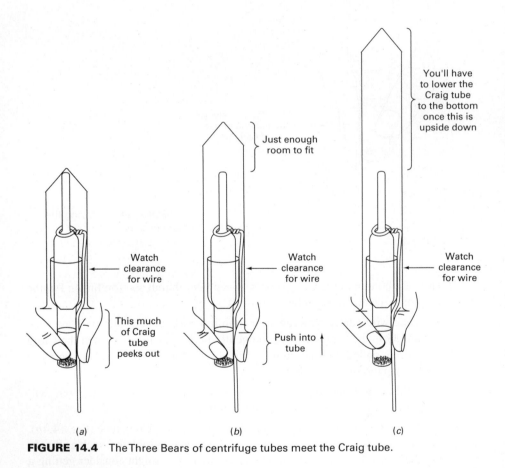

FIGURE 14.4 The Three Bears of centrifuge tubes meet the Craig tube.

CENTRIFUGING THE CRAIG TUBE

Just a few things here:

1. "Where does the centrifuge go when it takes a walk?" The centrifuge must be *balanced*—an even number of tubes that have almost identical weights, set in positions opposite each other.

2. If you've cut the copper wire properly, it should stick out only a little bit from the centrifuge tube. You wouldn't want to be flailed by a foot-long copper wire, would you? If necessary, bend it out of the way. Don't bend it so that you lose the wire into the centrifuge tube and can't get it again.

3. If centrifuging 1 minute is good, centrifuging 15 minutes is *NOT* better. Whatever the directions call for, just do it. Your yield will not improve. Trust me.

Getting the Crystals Out

After you've centrifuged the Craig tube and the centrifuge has stopped turning,

Do not stop the centrifuge with your fingers! Patience!

Pull the centrifuge tubes out, and carefully lift out the Craig tube. The solvent will have filled the bottom of the centrifuge tube, and the crystals will be packed down near the Craig tube top (Fig. 14.5).

I'd get a glass plate or watch glass, open the Craig tube, and put both sections on the glass to dry. (**CAUTION**—rolling Craig tubes will spread your product on the bench.)

After a bit, you can scrape your crystals into the appropriate tared container. (See "Tare to the Analytical Balance" in Chapter 5, "Microscale Jointware.") Don't try to get all of them; it's usually not worth the effort.

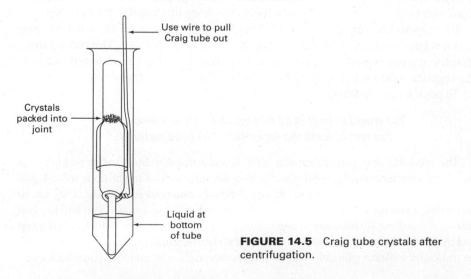

Use wire to pull Craig tube out

Crystals packed into joint

Liquid at bottom of tube

FIGURE 14.5 Craig tube crystals after centrifugation.

EXTRACTION AND WASHING

- *Add a few drops of a layer to 1 mL of water in a test tube: If it dissolves, it's an aqueous layer; if not, it's organic.*

- *Burp your funnels.*

- *Prevent vacuum buildup. Take the top stopper out when you empty a funnel.*

- *No solids in the funnel, EVER!*

Extraction is one of the more complex operations that you'll do in the organic chemistry lab. For this reason, I'll go over it especially slowly and carefully. Another term that you'll see used simultaneously with **extraction** is **washing**. That's because extraction and washing are really the same operation, but each leads to a different end. How else to put this?

Let's make some soup. Put the vegetables, fresh from the store, in a pot. You run cold water in and over them to clean them and throw this water down the drain. Later, you run water in and over them to cook them. You keep this water—it's the soup.

Both operations are similar. Vegetables in a pot in contact with water the first time is a **wash**. You remove unwanted dirt. *You washed with water.* The second time, vegetables in a pot in contact with water is an **extraction**. You've **extracted** essences of the vegetables *into water.* Very similar operations; very different ends.

To put it a little differently,

You would extract good material from an impure matrix.
You would wash the impurities from good material.

The vegetable soup preparation is a *solid–liquid extraction.* So is coffee making. You extract some components of a solid directly into the solvent. You might do a solid–liquid extraction in lab as a separate experiment; *liquid–liquid extractions are routine.* They are so common that if you are told to do an extraction or a washing, *it is assumed* you will use *two liquids—two insoluble liquids—*and a separatory funnel. The separatory funnel, called a **sep funnel** by those in the know, is a special funnel that you can separate liquids in. You might look at the section on separatory funnels (later in this chapter) right now, and then come back later.

Two insoluble liquids in a separatory funnel will form **layers**; one liquid will float on top of the other. You usually have compounds dissolved in these layers, and either the compound you want is *extracted from one to the other,* or junk you don't want is *washed from one layer to the other.*

Making the soup, you have *no* difficulty deciding what to keep or what to throw away. First you throw the water away; later you keep it. But this can change. In a sep funnel, the layer that you want to keep one time may not be the layer that you want to keep the next time. Yet, if you throw one layer away prematurely, you are doomed.

NEVER-EVER LAND

Never, never, never, never,
ever throw away any layer until you are absolutely sure you'll
never need it again. Not very much of your product can be
recovered from the sink trap!

I'm using a word processor, so I can copy this warning over and over again, but let's not get carried away. One more time. Wake Up Out There!

Never, never, never, never,
ever throw away any layer until you are absolutely sure you'll
never need it again. Not very much of your product can be
recovered from the sink trap!

STARTING AN EXTRACTION

To do any extraction, you'll need two liquids or solutions. *They must be insoluble in each other.* **Insoluble** here has a practical definition:

When mixed together, the two liquids form two layers.

One liquid will float on top of the other. A good example is ether and water. Handbooks say that ether is slightly soluble in water. When ether and water are mixed, yes, some of the ether dissolves; most of the ether just floats on top of the water.

Really soluble or *miscible liquid pairs are no good* for extraction and washing. When you mix them, *they will not form two layers!* In fact, they'll *mix in all proportions.* A good example of this is acetone and water. What kinds of problems can this cause? Well, for one, *you cannot perform any extraction with two liquids that are miscible.*

Let's try it. A mixture of, say, some mineral acid (is HCl all right?) and an organic liquid, "compound A," needs to have that acid washed out of it. You dissolve the compound A–acid mixture in some acetone. It goes into the sep funnel, and now you add water to wash out the acid.

Acetone is miscible in water. No layers form! You lose!

Back to the lab bench. Empty the funnel. Start over. This time, having called yourself several colorful names because you should have read this section thoroughly in the first place, you dissolve the compound A–acid mixture in ether and put it into the sep funnel. Add water, and *two layers form!* Now you can wash the acid from the organic layer to the water layer. The water layer can be thrown away.

Note that the acid went into the water, *then the water was thrown out!* So we call this a **wash**. If the water layer had been saved, we'd say the acid had been **extracted into the water layer**. It may not make sense, but that's how it is.

Review:

1. You *must have two insoluble liquid layers* to perform an extraction.

2. *Solids must be dissolved in a solvent,* and that solvent must be insoluble in the other extracting or washing liquid.

3. If you are washing or extracting an organic liquid, dissolve it into another liquid, *just like a solid,* before extracting or washing it.

So these terms, **extraction** and **washing**, are related. Here are a few examples.

1. Extract with ether. Throw ether together with the solution of product, and pull out *only the product into the ether.*

2. Wash with 10% NaOH. Throw 10% NaOH together with the solution of product, and pull out *everything but product into the NaOH.*

3. You can even extract with 10% NaOH.

4. You can even wash with ether.

So extraction is pulling out what you want from all else!
Washing is pulling out all else from what you want.

And please note—*you always do the pulling from one layer into another.* That's also *two immiscible liquids.*

You'll actually have to do a few of these things before you get the hang of it, but bear with me. When your head stops hurting, reread this section.

DUTCH UNCLE ADVICE

Before I go on to the separatory funnel, I'd like to comment on a few questions I keep hearing when people do washings and extractions.

1. *"Which layer is the water layer?"* This is so simple, it confounds everyone. Get a small (10 × 75-mm) test tube and add about 0.5–1.0 mL water. Now add 2–4 drops of the layer you *think* is aqueous, and swirl the tube. If the stuff *doesn't dissolve* in the water, it's *not* an aqueous (water) layer. The stuff may sink to the bottom, float on the top, do *both,* or even *turn the water cloudy!* It will *not,* however, dissolve.

2. *"How come I got three layers?"* Sometimes, when you pour fresh water or some other solvent into the funnel, you get a small amount hanging at the top, and it looks like there are three different layers. Yes, it *looks* as if there are three different layers, but there are not three different layers, only two layers, where part of one has lost its way. Usually this mysterious third layer looks just like its parent, and you might gently swirl the funnel and its contents to reunite the family.

3. *"What's the density of sodium hydroxide?"* You've just done a wash with 5–10% sodium hydroxide solution, you've just read something about finding various layers in the funnel by their densities, and, by this question, you've just shown that you've missed the point. Most wash solutions are 5–10% active ingredient dissolved in water. This means they are 90–95% water. Looking up the density of the solid reagents, then, is a waste of time. The densities of these solutions are *very close* to that of water. (10% NaOH has a specific gravity of 1.1089.)

4. *"I've washed this organic compound six times with sodium bicarbonate solution, so why's it not basic yet?"* This involves finding the pH of the organic layer. I'll give it away right now: *You cannot find the pH of an organic layer.* Not directly. You find the pH of the aqueous layer that's been in contact with the organic layer. If the aqueous layer is on the top, dip a glass rod into it and touch the glass rod to your test paper. If the aqueous layer is on the bottom and your sep funnel is in a ring, let a drop of the aqueous layer out of the funnel to hang on the outlet tip. Transfer the drop to your test paper.

Do not ever dip the test paper into any solution. Ever.

Warning. Be *sure* you are testing the aqueous layer. Some organics are very tenacious and can get onto your glass rod. The organic layer may wet the test paper, but without water, any color you see doesn't mean much.

THE SEPARATORY FUNNEL

Before we go on to some practical examples, you might want to know more about where all this washing and extracting is carried out. I've mentioned that it's a special funnel called a **separatory funnel** (Fig. 15.1) and that you can impress your friends by calling it a **sep funnel**. Here are a few things you should know.

The Stopper

At the top of the sep funnel should be a stopper, one of two kinds: either a plastic stopper with a single number on it, commonly 22 or 16, or a *standard taper glass stopper* commonly 19/22 marked on the stopper head (see further Chapter 4, "Jointware"). Make sure this number is on the head, and that it is the same as the number marked on the funnel. If this stopper is not so marked, you may find your product leaking over your shoes when you turn the sep funnel upside down.

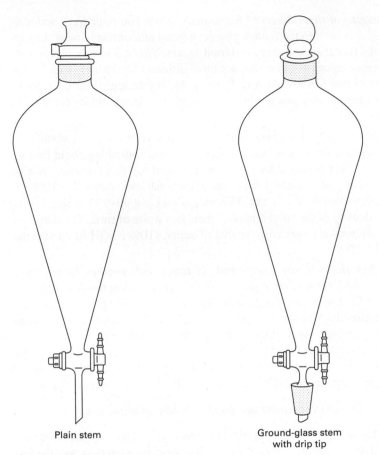

Plain stem

Ground-glass stem
with drip tip

FIGURE 15.1 Garden-variety separatory funnels.

The plastic stopper is part-and-parcel of an ordinary, plain-stem separatory funnel, will not fit standard taper joints, and does not require any greasing to keep it from sticking when you use the sep funnel. The glass stopper fits the standard taper joint on the addition funnel (and all of your other standard taper equipment as well), won't fit the plain-stem funnel, and might require grease to keep from sticking. If you use a funnel to pour liquids into the sep funnel, you'll keep that joint clean and probably won't need to use any grease. And if you do wind up greasing that joint, using the funnel should keep the grease out of your liquids.

Consult your instructor!

The Teflon Stopcock

In wide use today, the **Teflon stopcock** (Fig. 15.2) requires no grease and will not freeze up! A Teflon washer, a rubber ring, and, finally, a Teflon nut are placed on

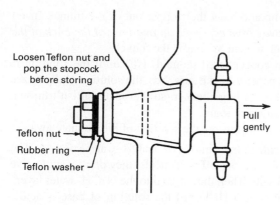

Loosen Teflon nut and
pop the stopcock
before storing

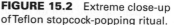

Pull
gently

Teflon nut

Rubber ring

Teflon washer

FIGURE 15.2 Extreme close-up
of Teflon stopcock-popping ritual.

the threads of the stopcock. This nut holds the whole thing on. Any leakage at this stopcock results from:

1. A loose Teflon nut. Tighten it.
2. A missing Teflon washer or rubber ring. Replace it.
3. An attempt to place the wrong size or wrong taper Teflon stopcock into the funnel. This is extremely rare. Get a new funnel.

Emergency stopcock warning!

Teflon may not stick, *but it sure can flow!* If the stopcock is extremely tight, the Teflon will bond itself to all the nooks and crannies in the glass in interesting ways. When you're through, always loosen the Teflon nut and "pop the stopcock" by pulling on the handle. The stopcock should be loose enough to spin freely when spun with one finger—*just remember to tighten it again before you use it.*

It seems to me that I'm the only one who reads the little plastic bags that hold the stopcock parts. Right on the bags, it shows that after the stopcock goes in, *the Teflon washer goes on the stem first,* followed by the rubber ring, and then the Teflon nut (Fig. 15.2). So why do I find most of these things put together incorrectly?

HOW TO EXTRACT AND WASH WHAT

Here are some practical examples of washings and extractions, covering various types and mixtures and separations, broken down into the four classifications listed previously.

1. *A strong organic acid.* Extract into saturated (sat'd) sodium bicarbonate solution. **CAUTION!** Foaming and fizzing and spitting and all sorts of carrying on. The weak base turns the strong acid into a salt, and the salt dissolves in the water–bicarbonate solution. Because of all the fizzing, you'll have to be very

careful. Pressure can build up and blow the stopper out of the funnel. Invert the funnel. *Point the stem away from everyone up and toward the back of the hood*—and open the stopcock to vent or "burp" the funnel.

   a. *To recover the acid,* add concentrated (conc'd) HCl until the solution is acidic. Use pH or litmus paper to make sure. Yes, the solution really fizzes and bubbles. You should use a large beaker, so that material isn't thrown onto the floor if there's too much foam.

   b. *To wash out the strong acid,* just throw the solution of bicarbonate away.

2. **A weakly acidic organic acid.** Extract into 10% NaOH–water solution. The strong base is needed to rip the protons off weak acids (they don't want to give them up) and turn them into salts. Then they'll go into the NaOH–water layer.

   a. *To recover the acid,* add conc'd HCl until the solution of base is acidic when tested with pH or litmus paper.

   b. *To wash out the weak acid,* just throw this NaOH–water solution away.

3. **An organic base.** Extract with 10% HCl–water solution. The strong acid turns the base into a salt. (This *turning the whatever into a salt that dissolves in the water solution* should be pretty familiar to you by now. Think about it.) Then the salt goes into the water layer.

   a. *To recover the base,* add ammonium hydroxide to the water solution until the solution is *basic* to pH or litmus paper. *Note that this is the reverse of the treatment given to organic acids.*

   b. *To wash out an organic base, or any base,* wash as previously noted and throw out the solution.

4. **A neutral organic.** If you've extracted strong acids first, then weak acids, and then bases, there are only neutral compound(s) left. If possible, just remove the solvent that now contains *only your neutral compound.* If you have *more than one neutral compound,* you may want to extract one from the other(s). You'll have to find *two different immiscible organic liquids,* and *one liquid must dissolve only the neutral organic compound you want!* A tall order. You must count on *one neutral organic compound being more soluble in one layer than in the other.* Usually the separation is *not clean—not complete.* And you have to do more work.

What's "more work"? That depends on the results of your extraction.

The Road to Recovery—Back-Extraction

I've mentioned **recovery** of the four types of extractable materials, but that's not all the work you'll have to do to get the compounds in shape for further use.

1. If the recovered material is *soluble in the aqueous recovery solution,* you'll have to do a **back-extraction**.

   a. Find a solvent that dissolves your compound and is *not miscible in the aqueous recovery solution.* This solvent should boil at a low temperature (<100°C), since you will have to remove it. Ethyl ether is a common choice. (**Hazard!** *Very flammable.*)

b. *Then you extract your compound back from the aqueous recovery solution into this organic solvent.*

c. Dry this organic solution with a drying agent (see Chapter 10, "Drying Agents").

d. Now you can remove the organic solvent. Either distill the mixture or evaporate it, perhaps on a steam bath. All this is done away from flames and in a hood.

When you're through removing the solvent and your product is not pure, clean it up. If your product is a liquid, you might distill it; if it is a solid, you might recrystallize it. Make sure it is clean.

2. If the recovered material is *insoluble in the aqueous recovery solution* and *it is a solid,* collect the crystals on a Buchner funnel. If they are *not pure,* you should recrystallize them.

3. If the recovered material is *insoluble in the aqueous recovery solution* and *it is a liquid,* you can use your separatory funnel to *separate the aqueous recovery solution from your liquid product. Then dry your liquid product and distill it if it is not clean. Or you might just do a back-extraction as just described.* This has the added advantage of removing the small amount of liquid product that dissolves in the aqueous recovery solution and increases your yield. Remember to dry the back-extracted solution before you remove the organic solvent. Then distill your liquid compound if it is not clean.

A Sample Extraction

I think the only way I can bring this out is to use a typical example. This may ruin a few lab quizzes, but if it helps, it helps.

Say you have to separate a mixture of *benzoic acid (1), phenol (2), p-toluidine (4-methylaniline) (3),* and *anisole (methoxybenzene) (4)* by extraction. The numbers refer to the class of compound, as previously listed. We're assuming that none of the compounds reacts with any of the others and that you know we're using all four types as indicated. Phenol and 4-methylaniline are corrosive toxic poisons, and if you get near these compounds in lab, *be very careful.* When they are used as an example on these pages, however, you are quite safe. Here's a sequence of tactics.

1. Dissolve the mixture in ether. Ether is insoluble in the water solutions you will extract into. Ether happens to dissolve all four compounds. Aren't you lucky? You bet! It takes lots of hard work to come up with the "typical student example."

2. Extract the ether solution with 10% HCl. This converts *only* compound 3, the basic *p*-toluidine, into the hydrochloride salt, which dissolves in the 10% HCl layer. You have just *extracted a base with an acid solution.* Save this solution for later.

3. Now extract the ether solution with sat'd sodium bicarbonate solution. *Careful!* Boy, will this fizz! Remember to swirl the contents and release the pressure. The weak base converts *only* compound 1, the *benzoic acid, to a salt, which dissolves in the sat'd bicarbonate solution.* Save this for later.

4. Now extract the ether solution with the 10% NaOH solution. This converts compound 2, *weak acid,* phenol, to a salt, which dissolves in the 10% NaOH layer. Save this for later. *If you do this step before step 3, that is, extract with 10% NaOH solution before the sodium bicarbonate solution, both the weak acid,* phenol, *and the strong acid,* benzoic acid, *will be pulled out into the sodium hydroxide.* Ha-ha. This is the usual kicker they put in lab quizzes, and people always forget it.

5. The only thing left is the neutral organic compound dissolved in the ether. Just drain this solution into a flask.

So, now we have four flasks with four solutions with one component in each. *They are separated.* You may ask, "How do we get these back?"

1. *The basic compound (3).* Add ammonium hydroxide until the solution turns basic (test with litmus or pH paper). The *p*-toluidine, or organic base (3), is regenerated.

2. *The strong acid or the weak acid (1, 2).* A bonus. Add diluted HCl until the solution turns acidic to an indicator paper. Do it to the other solution. Both acids will be regenerated.

3. *The neutral compound (4).* It's in the ether. If you evaporate the ether (*no flames!*), the compound should come back.

Now, when you recover these compounds, sometimes they don't come back in such good shape. You will have to do more work.

1. Addition of HCl to the benzoic acid extract will produce huge amounts of white crystals. Get out the Buchner funnel and have a field day! Collect all you want. But they won't be in the best of shape. Recrystallize them. (**Note:** *This compound is insoluble in the aqueous recovery solution.*)

2. The phenol extract is a different matter. You see, *phenol is soluble in water,* and it doesn't come back well at all. So, get some fresh ether, extract the phenol from HCl solution to the ether, and evaporate the ether. Sounds crazy, no? No. Remember, I called this a **back-extraction**, and you'll have to do this more often than you would like to believe. (**Note:** *This compound is soluble in the aqueous recovery solution.*)

3. The *p*-toluidine should return after the addition of ammonium hydroxide. Recrystallize it from ethanol, so that it also looks respectable again.

4. The neutral anisole happens to be a *liquid (bp 155°C),* and you'll have to take care when you evaporate the ether that you don't lose much anisole. Of course, you shouldn't expect to see any crystals. Now this neutral anisole liquid that comes back after you've evaporated the ether (*no flames!*) will probably be contaminated with a little bit of all of the other compounds that started out in the ether. You will have to purify this liquid, probably by a simple distillation.

You may or may not have to do all of this with the other solutions or with any other solution that you ever extract in your life. You must choose. Art over science. As confusing as this is, I have simplified it a lot. Usually you have to extract these solutions more than once, and the separation is not as clean as you'd like. Not 100% but pretty good. If you are still confused, see your instructor.

Performing an Extraction or Washing

1. Suspend a sep funnel in an iron ring.

2. Remove the stopper.

3. *Make sure the stopcock is closed!* You don't really want to scrape your product off the bench top.

4. Add the solution to be extracted or washed. Less than half full, please. Add the extraction or washing solvent. An equal volume is usually enough. The funnel is funnel shaped, and the equal volumes won't look equal.

5. Replace the stopper.

6. Remove the sep funnel from the iron ring. Hold the stopper and stopcock tightly. Pressure may build up during the next step and blow your product out onto the floor.

7. Invert the sep funnel (Fig. 15.3).

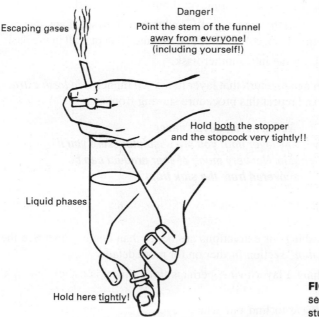

Escaping gases

Danger!
Point the stem of the funnel
<u>away from everyone!</u>
(including yourself!)

Hold <u>both</u> the stopper
and the stopcock very tightly!!

Liquid phases

Hold here <u>tightly!</u>

FIGURE 15.3 Holding a sep funnel so as not to get stuff all over.

> *Point the stem up away from everyone—*
> *up into the back of a hood if at all possible!*

Make *sure* the liquid has drained down away from the stopcock, then *slowly* open the stopcock. You may hear a whoosh, possibly a pfffft, as the pressure is released. This is due to the high vapor pressure of some solvents or to a gas evolved from a reaction during the mixing. This can cause big trouble when you are told to neutralize acid by washing with sodium carbonate or sodium bicarbonate solutions.

8. Close the stopcock!

9. Shake the funnel gently, invert it, and open the stopcock again.

10. Repeat steps 8 and 9 until no more gas escapes.

11. If you see that you might get an **emulsion**—*a fog of particles*—with this gentle inversion, *do NOT shake the funnel vigorously.* You might have to continue the rocking and inverting motions 30 to 100 times, as needed, to get a separation. Check with your instructor and with the hints on breaking up emulsions (see "Extraction Hints," following). Otherwise, shake the funnel vigorously about 10 times to get good distribution of the solvents and solutes. Really shake it.

12. Put the sep funnel back in the iron ring.

13. *Remove the glass stopper.* Otherwise the funnel won't drain, and you'll waste your time just standing there.

14. Open the stopcock, and let the bottom layer drain off into a flask.

15. Close the stopcock, swirl the funnel gently, and then wait to see if any more of the bottom layer forms. If so, collect it. If not, assume you got it all in the flask.

16. Let the remaining layer out into another flask.

To extract any layer again, return that layer to the sep funnel, *add fresh extraction or washing solvent,* and repeat this procedure starting from step 5.

> *Never, never, never, never,*
> *ever throw away any layer until you are absolutely sure you'll*
> *never need it again. Not very much of your product can be*
> *recovered from the sink trap!*

Extraction Hints

1. Several smaller washings or extractions are better than one big one. See the *"Theory of Extraction"* section further on for more details.

2. Extracting or washing a layer *twice,* perhaps three times, is usually enough. Diminishing returns set in after that.

3. Sometimes you'll have to find out which layer is **the water layer**. This is so simple, it confounds everyone. Add 2–4 drops of each layer to a test tube containing 1 mL of water. Shake the tube. If the stuff *doesn't dissolve* in the

water, it's *not* an aqueous (water) layer. The stuff may sink to the bottom, float on the top, do *both,* or even *turn the water cloudy!* It will *not,* however, dissolve.

4. If *only the top layer* is being extracted or washed, *it does not have to be removed from the funnel,* ever. Just drain off the bottom layer, and then add more *fresh* extraction or washing solvent. Ask your instructor about this.

5. You *can* combine the extracts of a multiple extraction, *if they have the same material in them.*

6. If you have to wash your organic compound with water and the organic is *slightly soluble in water,* try washing with *saturated salt solution.* The theory is that if all that salt dissolved in the water, what room is there for your organic product? This point is a favorite of quizmakers and should be remembered. It's the same thing that happens when you add salt to reduce the solubility of your compound during a crystallization (see "Salting Out" in Chapter 13).

7. If you get an **emulsion,** you have not two distinct layers, but a kind of a *fog of particles.* Sometimes you can break up the charge on the suspended droplets by adding a little salt or some acid or base. Or add ethanol. Or stir the solutions slowly with a glass rod. Or gravity-filter the entire contents of your separatory funnel through filter paper. Or laugh. Or cry. Emulsion-breaking is a bit of an art. Careful with the acids and bases, though. They can react with your product and destroy it.

8. If you decide to add salt to a sep funnel, don't add so much that it clogs up the stopcock! For the same reason, keep drying agents out of sep funnels.

9. Sometimes some material comes out, or it will not dissolve in the two liquid layers and hangs in there at the **interface**. It may be that there's not enough liquid to dissolve this material. One cure is to *add more fresh solvent of one layer or the other.* The solid may dissolve. If there's no room to add more, you may have to remove *both* (yes, both) layers from the funnel, and try to dissolve this solid in either of the solvents. It can be confusing. If the material does *not* redissolve, then it is a new compound and should be saved for analysis. You should see your instructor for that one.

Theory of Extraction

"Several small extractions are better than one big one." Doubtless you've heard this many times, but now I'm going to try to show that it is true.

By way of example, let's say you have an *aqueous* solution of oxalic acid, and you need to isolate it from the water by doing an extraction. Looking them up, you find some solubilities of oxalic acid as follows: 9.5 g/100 g in water; 23.7 g/100 g in ethanol; 16.9 g/100 g in diethyl ether. Based on the solubilities, you decide to extract into ethanol from water, forgetting for the moment that ethanol is *soluble in water* and that you *must have two insoluble liquids* to carry out an extraction. Chagrined, you forget the ethanol and choose diethyl ether.

From the preceding solubility data, we can calculate the **distribution**, or **partition coefficient**, for oxalic acid in the water–ether extraction. This coefficient (number) is just the ratio of solubilities of the compound you wish to extract in the two layers. Here,

$$K_p = \frac{\text{solubility of oxalic acid in ether}}{\text{solubility of oxalic acid in water}}$$

which amounts to 16.9/9.5, or 1.779.

Imagine that you have 40 g of oxalic acid in 1000 mL water and you put that in contact with 1000 mL ether. The oxalic acid *distributes itself between the two layers*. How much is left in each layer? Well, if we let x g equal the amount that stays in the water, $1.779x$ g of the acid has to walk over to the ether. And so

Wt of oxalic acid in ether = (1000 mL)(1.779x g/mL) = 1779x g

Wt of oxalic acid in water = (1000 mL)(x g/mL) = 1000x g

The total weight of the acid is 40 g (now partitioned between two layers) and

$$2779x \text{ g} = 40 \text{ g}$$
$$x = 0.0144$$

and

Wt of oxalic acid in ether = 1779(0.0144) g = 25.6 g

Wt of oxalic acid in water = 1000(0.0144) g = 14.4 g

Now, let's start with the same 40 g of oxalic acid in 1000 mL of water, but this time we will *do three extractions with 300 mL of ether*. The first 300-mL portion hits, and

Wt of oxalic acid in ether = (300 mL)(1.779x g/mL) = 533.7x g

Wt of oxalic acid in water = (1000 mL)(x g/mL) = 1000x g

The total weight of the acid is 40 g (now partitioned between two layers) and

$$1533.7x \text{ g} = 40 \text{ g}$$
$$x = 0.0261$$

so

Wt of oxalic acid in ether = 533.7(0.0261) g = 13.9 g

Wt of oxalic acid in water = 1000(0.0261) g = 26.1 g

That ether layer is removed, and the *second jolt* of 300 mL fresh ether hits, and

Wt of oxalic acid in ether = (300 mL)(1.779x g/mL) = 533.7x g

Wt of oxalic acid in water = (1000 mL)(x g/mL) = 1000x g

But here, we started with 26.1 g of acid in water (now partitioned between two layers) and

$$1533.7x \text{ g} = 26.1 \text{ g}$$
$$x = 0.0170$$

so

$$\text{Wt of oxalic acid in ether} = 533.7(0.0170) \text{ g} = 9.1 \text{ g}$$
$$\text{Wt of oxalic acid in water} = 1000(0.0170) \text{ g} = 17.0 \text{ g}$$

Again, *that ether layer is removed*, and the *third jolt* of 300 mL fresh ether hits, and

$$\text{Wt of oxalic acid in ether} = (300 \text{ mL})(1.779x \text{ g/mL}) = 533.7x \text{ g}$$
$$\text{Wt of oxalic acid in water} = (1000 \text{ mL})(x \text{ g/mL}) = 1000x \text{ g}$$

But here, we started with 17.0 g of acid in water (now partitioned between two layers) and

$$1533.7x \text{ g} = 17.0 \text{ g}$$
$$x = 0.011$$

so

$$\text{Wt of oxalic acid in ether} = 533.7(0.011) \text{ g} = 5.87 \text{ g}$$
$$\text{Wt of oxalic acid in water} = 1000(0.011) \text{ g} = 11.0 \text{ g}$$

(They don't quite add up to 17.0 g—I've rounded them off a bit.)

Let's consolidate what we have: first, 13.9 g, then 9.1 g, and finally 5.87 g of oxalic acid, for a total of 28.9 g (72.3%) of acid extracted into 900 mL of ether. OK, that's not far from 25.6 g (64%) extracted *once* into 1000 mL of ether. That's because the distribution coefficient is fairly low. But it *is* more. That's because *several small extractions are better than one large one*.

EXERCISES

1. What are the characteristics of a good extraction solvent?

2. In a typical "Extraction of Caffeine" experiment, you use boiling water to get the caffeine out of the coffee or tea, and then extract this aqueous solution with diethyl ether or methylene chloride. Why can't you immediately put the aqueous solution into the sep funnel with the ether or methylene chloride in order to save time?

3. Compare and contrast extraction with washing.

4. The Teflon stopcock of separatory funnels is kept on the funnel by a Teflon nut, a Teflon washer, and a rubber ring. What is the correct order, starting from the glass? Do your separatory funnels have these things on in the correct order?

EXTRACTION AND WASHING: Microscale

- *Guard against vapor pressure buildup in pipets; your product might blow out.*

In microscale, instead of a sep funnel you use a conical vial and some Pasteur pipets. First you mix your extraction solvent with your product; then you separate the two liquids. And if you don't see any drawings with the pipet and bulb upside down, that's because you're NOT supposed to do that!

MIXING

1. Put the material to be extracted into an appropriately sized conical vial. This vial should be, at *minimum*, twice the volume of the liquid you want to extract. Usually, this conical vial is the reaction vial for the experiment, so the choice is easily made (forced on you).

2. Add an appropriate solvent (*not*—repeat—*not* extracting liquid), so the volume of the mixture is about 1 mL. (What if it's about 1 mL already?)

3. Add about 1 mL of extracting solvent to the vial.

4. At this point, cap the vial and shake to mix (**Caution**—leaks!). Better you should add a **magnetic spinning vane** and spin the two layers for a minute or so.

SEPARATION: REMOVING THE BOTTOM LAYER (FIG. 16.1)

1. Have an empty vial ready.

2. Prewet a Pasteur filter pipet and bulb combo with the extraction solvent. (See Chapter 7, "Pipet Tips.")

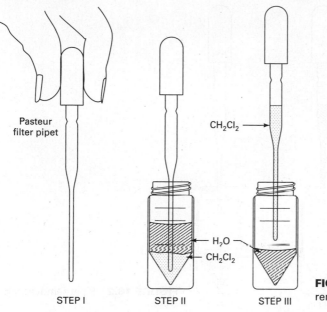

Pasteur
filter pipet

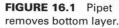
CH_2Cl_2

H_2O
CH_2Cl_2

STEP I   STEP II   STEP III

FIGURE 16.1 Pipet removes bottom layer.

3. Squeeze the bulb.

4. Stick the pipet into the vial down to the bottom.

5. Slowly—read that word again—slowly, draw the bottom layer up into the pipet. Get it all now. (Actually, you could do this a few times if necessary.) And don't suck any liquid *into* the rubber bulb!

6. Squirt the bottom layer into an empty vial.

SEPARATION: REMOVING THE TOP LAYER (FIG. 16.2)

1. Have an empty vial ready.

2. Prewet a Pasteur filter pipet and bulb combo with the extraction solvent. (See Chapter 7, "Pipet Tips.")

3. Really squeeze the bulb.

4. Stick the pipet into the vial down to the bottom.

5. Slowly—read that word again—slowly, draw *all*—yes *all,* I mean both layers—into the pipet. *Don't slurp!* If you draw lots of air into the pipet *after* both layers are in there, the air bubble will mix both layers in the pipet. Even worse, because of evaporation into this new "fresh" air, vapor pressure buildup can make both layers fly out back into the vial, onto the bench top, on to your hands, wherever. And don't suck any of this *into* the rubber bulb!

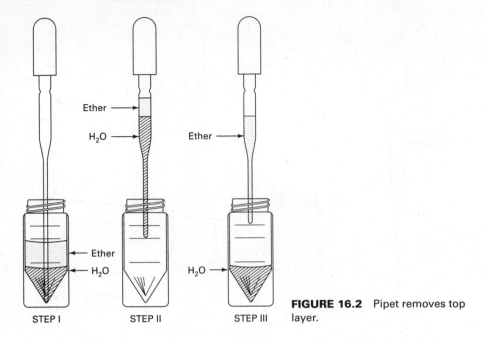

Ether →

H₂O →

Ether →

← Ether

← H₂O

H₂O →

STEP I

STEP II

STEP III

FIGURE 16.2 Pipet removes top layer.

6. Squeeze the bulb to put the *bottom* layer back into the original vial.

7. Squirt the top layer into an empty vial.

SEPARATION: REMOVING *BOTH* LAYERS

Some say, if you already have both layers in the pipet (Fig. 16.2), why bother squirting the lower layer, or any layer really, back into the vial? Suck 'em both up, and squirt each into whatever vessel you need to. "Squirt" slowly, gently. Remember, you're separating two 1-mL portions of liquid here. And it's probably better to leave a little of the organic layer in with the water layer here. Drying 1 mL of a liquid is going to be tricky enough without extra water from the water layer in there.

Chapter 17

SOURCES OF HEAT

- Burners are the last resort.
- Hot glassware looks a lot like cold glassware. Careful!

Many times you'll have to heat something. Don't just reach for the Bunsen burner. That flame you start just may be your own. There are alternative sources you should think of *first*. And **Hot glassware can look a lot like cold glassware**.

BOILING STONES

All you want to do is start a distillation. Instructor walks up and says,

"Use a boiling stone or it'll bump."

"But I'm only gonna . . ."

"Use a boiling stone or it'll bump."

"It's started already and . . ."

"Use a boiling stone or it'll bump."

"I'm not gonna go and . . ."

Suddenly—whoosh! Product all over the bench! Next time, you put a boiling stone in *before* you start. No bumping. But your instructor won't let you forget the time you did it your way.

Don't let this happen to you. Use a **brand-new boiling stone** every time you have to boil a liquid. A close-up comparison between a boiling stone and the inner walls of a typical glass vessel reveals thousands of tiny nucleating points on the stone where vaporization can take place, in contrast to the smooth glass surface that can hide unsightly hot spots and lead to bumping, a massive instantaneous vaporization that will throw your product all over.

CAUTION! Introducing a boiling stone into hot liquid may result in instant vaporization and loss of product. Remove the heat source, swirl the liquid to remove hot spots, and then add the boiling stone.

Used as directed, the boiling stone will relieve minor hot spots and prevent loss of product through bumping. So remember . . . whenever you boil, wherever *you boil,*

Always use a fresh boiling stone!

Don't be the last on your bench to get this miracle of modern science made exclusively from nature's most common elements.

THE STEAM BATH

If one of the components boils below 70°C and you use a *Bunsen burner,* you may have a hard time putting out the fire. Use a **steam bath** (Fig. 17.1)!

1. Find a steam tap. It's like a water tap, only this one dispenses steam. **CAUTION!** You can get burned.

2. Connect tubing to the tap *now.* It's going to get awfully hot in use. Make sure you've connected a piece that'll be long enough to reach your steam bath.

3. Don't connect this tube to the steam bath yet! Just put it into a sink. Because steam lines are usually full of water from condensed steam, *drain the lines first;* otherwise, you'll waterlog your steam bath.

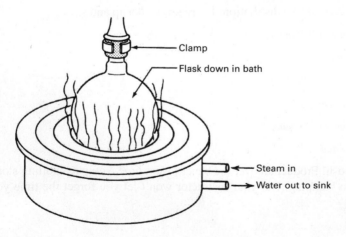

FIGURE 17.1 The steam bath in use.

4. **CAUTION!** Slowly open the steam tap. You'll probably hear bonking and clanging as steam enters the line. Water will come out. It'll get hotter and may start to spit.

5. Wait until the line is mostly clear of water, then turn off the steam tap. *Wait for the tubing to cool.*

6. Slowly, carefully, and cautiously, *making sure the tube is not hot,* connect the tube to the inlet of the steam bath. This is the *uppermost* connection on the steam bath.

7. Connect another tube to the outlet of the steam bath—the *lower* connection—and to a drain. Any water that condenses in the bath while you're using it will drain out. Make sure all parts of the drain hose are *lower* than the bath to keep the water flowing out of the bath.

Usually, steam baths have concentric rings as covers (Fig. 17.1). You can control the "size" of the bath by removing various rings.

Never do this after you've started the steam. You will get burned!

And don't forget— round-bottom flasks should be about halfway in the bath. You shouldn't really let steam rise up all around the flask. Lots of steam will certainly steam up the lab and may expose you to corrosion inhibitors (morpholine) in the steam lines.

THE BUNSEN BURNER

The first time you get the urge to take out a Bunsen burner and light it up, *don't*. You may blow yourself up. Please check with your instructor to see if you even need a burner. **Once you find out you *can* use a burner**, assume that the person who used it last didn't know much about burners. Take some precautions so as not to burn your eyebrows off.

Now, Bunsen burners are not the only kind. There are **Tirrill burners** and **Meker burners** as well. Some are fancier than others, but they work pretty much the same. So when I say **burner** anywhere in the text, it could be any of them.

1. Find the round knob that controls the **needle valve**. This is at the base of the burner. Turn it fully clockwise (inward) to stop the flow of gas completely. If your burner doesn't have a needle valve, it's a traditional Bunsen burner, and the gas flow has to be regulated at the bench stopcock (Fig. 17.2). This can be dangerous, especially if you have to reach over your apparatus and burner to turn off the gas. Try to get a different model.

2. There is a **movable collar** at the base of the burner which controls air flow. For now, see that *all* the holes are closed (i.e., no air gets in).

3. Connect the burner to the bench stopcock by some tubing and turn the bench valve *full on*. The bench valve handle should be parallel to the outlet (Fig. 17.2).

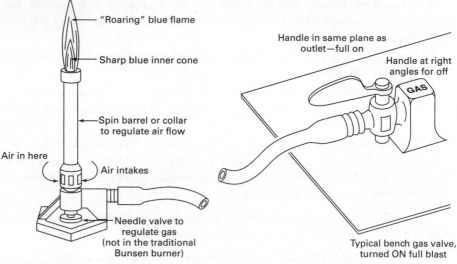

FIGURE 17.2 More than you care to know about burners.

4. Now, *slowly* open the needle valve. You may be just able to hear some gas escaping. Light the burner. *Mind your face!* Don't look down at the burner as you open the valve.

5. You'll get a wavy yellow flame, something you don't really want. But at least it's lit. Now open the air collar a little. The yellow disappears; a blue flame forms. This is what you want.

6. Now, adjust the needle valve and collar (the adjustments play off each other) for a steady blue flame.

Burner Hints

1. Air does not burn. You must wait until the gas has pushed the air out of the connecting tubing. Otherwise, you might conclude that none of the burners in the lab work. Patience, please.

2. When you set up for distillation or reflux, don't waste a lot of time *raising* and *lowering* the entire setup so that the burner will fit. This is nonsense. Move the burner! Tilt it (Fig. 17.3). If you leave the burner motionless under the flask, you may scorch the compound, and your precious product can become a "dark intractable material."

3. Placing a wire gauze between the flame and the flask spreads out the heat evenly. Even so, the burner may have to be moved around. Hot spots can cause star cracks to appear in the flask (see "Round-Bottom Flasks" in Chapter 4, "Jointware").

4. *Never* place the flask in the ring without a screen (Fig. 17.4). The iron ring heats up faster than the flask, and the flask cracks in the nicest line you've ever seen around it. The bottom falls off, and the material is all over your shoes.

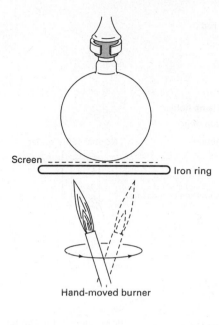

Screen

Iron ring

Hand-moved burner

FIGURE 17.3   Don't raise the flask, lower the burner.

FIGURE 17.4   Flask in the iron ring.

THE HEATING MANTLE

A very nice source of heat, the heating mantle takes some special equipment and finesse (Fig. 17.5).

1. *Variable-voltage transformer.* The transformer takes the quite lethal 120 V from the wall socket and can change it to an equally dangerous 0 to 120 V, depending on the setting on the dial. Unlike temperature settings on a Mel-Temp, on a transformer 0 means 0 V, 20 means 20 V, and so on. I like to start at 0 V and work my way up. Depending on how much heat you want, values from 40 to 70 seem to be good places to start. You'll also need a cord that can plug into both the transformer and the heating mantle. (See Chapter 12, "The Melting-Point Experiment.")

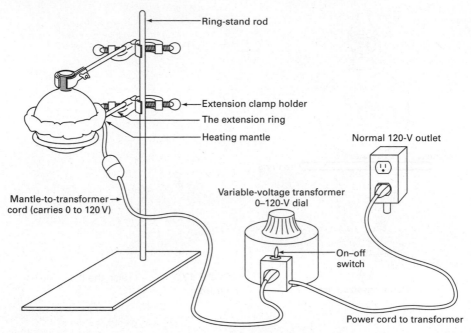

FIGURE 17.5 Round-bottom flask and mantle ready to go.

2. *Traditional fiberglass heating mantle.* An electric heater wrapped in fiberglass insulation and cloth that looks vaguely like a catcher's mitt (Fig. 17.5).

3. *Thermowell heating mantle.* You can think of the Thermowell heating mantle as the fiberglass heating mantle in a can. In addition, there is a hard ceramic shell that your flask fits into (Fig. 17.6). Besides just being more mechanically sound, it'll help stop corrosive liquids from damaging the heating element if your flask cracks while you're heating it.

FIGURE 17.6 A Thermowell heating mantle.

4. *Things not to do.*

   a. *Don't ever plug the mantle directly into the wall socket!* I know, the curved prongs on the mantle connection won't fit, but the straight prongs on the adapter cord will. Always use a variable-voltage transformer and start with the transformer set to zero.

   b. *Don't use too small a mantle.* The only cure for this is to *get one that fits properly.* The poor contact between the mantle and the glass doesn't transfer heat readily, and the mantle burns out.

   c. *Don't use too large a mantle.* The only good cure for this is to *get one that fits properly.* An acceptable fix is to *fill the mantle with sand after the flask is in* but before you turn the voltage on. Otherwise, the mantle will burn out.

Hint: When you set up a heating mantle to heat any flask, usually for **distillation** or **reflux**, put the mantle on an iron ring and keep it clamped a few inches above the bench top (Fig. 17.5). Then clamp the flask *at the neck* in case you have to remove the heat quickly. You can just unscrew the lower clamp and drop the mantle and iron ring.

PROPORTIONAL HEATERS AND STEPLESS CONTROLLERS

In all of these cases of heating liquids for distillation or reflux, we really control the *electric power* directly, not the heat or temperature. Power is applied to the heating elements, and they warm up. Yet the final temperature is determined by the heat loss to the room, the air, and, most important, the flask you're heating. There are several types of electric power controls.

1. *The variable-voltage transformer.* We've discussed this just previously. Let me briefly restate the case: Set the transformer to 50 on the 0–100 dial and you get 50% of the line voltage, all the time, night and day, rain or shine.

2. *The mechanical stepless controller.* This appears to be *the* inexpensive replacement for the variable-voltage transformer. Inside one model, there's a small heating wire wound around a bimetal strip with a magnet at one end (Fig. 17.7). A plunger connected to the dial on the front panel changes the distance between the magnet and a metal plate. When you turn the device on with a heating mantle attached, current goes through the small heating wire and the mantle. The mantle is now on *full blast* (120 V out of 120 V from the electric wall socket)! As the small heating wire warms the bimetal strip, the strip expands, distorts, and finally pulls the magnet from its metal plate, opening the circuit. The mantle now cools rapidly (0 V out of 120 V from the wall socket), along with the bimetal strip. Eventually, the strip cools enough to let the magnet get close to that metal plate, and—CLICK—everything's on full tilt again.

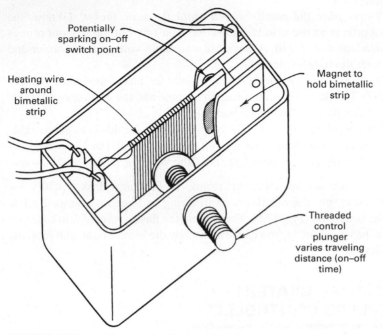

FIGURE 17.7 Inside a mechanical stepless heater.

The front panel control varies the **duty cycle**, the time the controller is *full on* to the time the controller is *full off*. If the flask, contents, and heating mantle are substantial, it takes a long time for them to warm up and cool down. A setup like that would have a large **thermal lag**. With small setups (approximately 50 mL or so), there is a small thermal lag, and very wild temperature fluctuations can occur. Also, operating a heating mantle this way is just like repeatedly plugging and unplugging it directly into the wall socket. Not many devices easily take that kind of treatment.

3. *The electronic stepless controller.* Would you believe a *light dimmer*? The electronic controller has a *triac,* a semiconductor device that lets fractions of the AC power cycle through to the heating mantle. The AC power varies like a sine wave, from 0 to 120 V from one peak to the next. At a setting of 25%, the triac remains off during much of the AC cycle, finally turning on when the time is right (Fig. 17.8). Although the triac does turn "full off and full on," it does so at times in so carefully controlled a way that the mantle never sees full line power (unless you deliberately set it there).

Only the mechanical controller can actually spark, and, under the right conditions of concentration of flammable vapors, set them afire. Might be useful knowing what's controlling your heating.

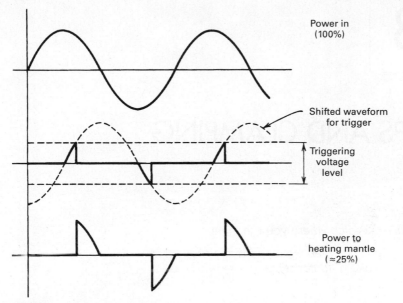

FIGURE 17.8   Light dimmer and heating mantle triac power control.

EXERCISE

Why shouldn't you plug a heating mantle directly into the electrical outlet on the bench?

Chapter 18

CLAMPS AND CLAMPING

- ■ *Plastic clips melt. Watch where you use them.*
- ■ *Clamps move up a bit when tightened to the rod.*
  Use this fact to make a tighter setup.

Unfortunately, glass apparatus needs to be held in place with more than just spit and baling wire. In fact, you would do well to use clamps. Life would be simple if there were just one type of fastener, but that's not the case.

1. ***The simple buret clamp*** (Fig. 18.1). Though it's popular in other chem labs, the simple buret clamp just doesn't cut it for organic lab. The clamp is too short, and adjusting angles with the locknut (by loosening the locknut, swiveling the clamp jaws to the correct angle, and tightening the locknut against the *back* stop, away from the jaws) is not a great deal of fun. If you're not careful, the jaws will slip right around and all the chemicals in your flask will fall out.

2. ***The simple extension clamp and clamp fastener*** (Fig. 18.2). This two-piece beast is the second-best clamp going. It is much longer (approximately 12 in.), so you can easily get to complex setups. By loosening the **clamp holder thumbscrew**, the clamp can be pulled out, pushed back, or rotated to any angle. By loosening the **ring-stand thumbscrew**, the clamp, along with the clamp holder, can move up and down.

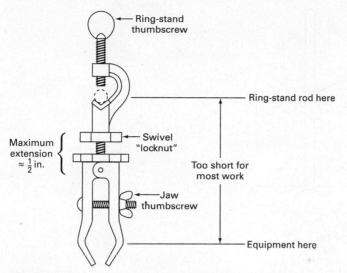

FIGURE 18.1 The "barely adequate for organic lab" buret clamp.

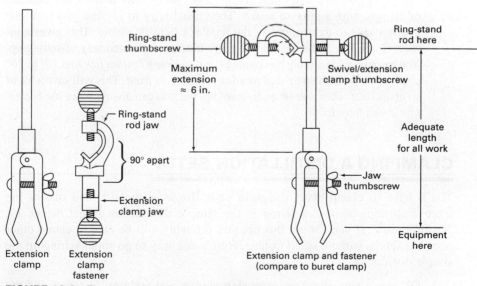

FIGURE 18.2 The extension clamp and clamp fastener.

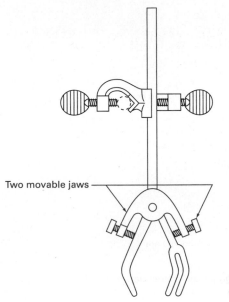

Two movable jaws

FIGURE 18.3 The three-fingered clamp with clamp fastener.

3. *The three-fingered extension clamp* (Fig. 18.3). This is truly the Cadillac of clamps, with a price to match. They usually try to confuse you with *two thumbscrews* to tighten, unlike the regular extension clamp. This gives a bit more flexibility, at the cost of a slightly more complicated way of setting up. You can make life simple by opening the *two-prong bottom jaw to a 10° to 20° angle from the horizontal and treating that jaw as fixed.* This will save a lot of wear and tear when you set equipment up, but you can *always move the bottom jaw* if you have to.

CLAMPING A DISTILLATION SETUP

You'll have to clamp many things in your life as a chemist, and one of the more frustrating setups to clamp is the **simple distillation** (see Chapter 19, "Distillation"). If you can set this up, you probably will be able to clamp other common setups without much trouble. Here's one way to go about setting up the simple distillation.

1. OK, get a **ring stand**, an **extension clamp**, and a **clamp fastener**, and put them all together. What heat source? A Bunsen burner, and you'll need more

room than you do with a heating mantle (see Chapter 17, "Sources of Heat"). In any case, you don't know where the receiving flask will show up, and then you might have to readjust the entire setup. Yes, you should have read the experiment before hand, so you'd know about the heating mantles.

2. Clamp the flask (around the neck) a few inches up the ring stand (Fig. 18.4). We *are* using heating mantles, and you'll need the room underneath to drop the mantle in case it gets too hot. That's why the flask is *clamped at the neck. Yes. That's where the flask is ALWAYS clamped, no matter what the heat source,* so it doesn't fall when the mantle comes down. What holds the mantle? Extension ring and clamp fastener.

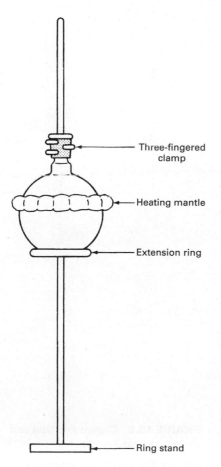

Three-fingered clamp

Heating mantle

Extension ring

Ring stand

FIGURE 18.4 Flask and heating mantle out on a ring stand.

3. Remember, whether you set these up from left to right, or right to left—*distilling flask first!*

4. Add the **three-way adapter** now (Fig. 18.5). Thermometer and thermometer adapter come later.

5. Put a second ring stand about one condenser length away from the first ring stand. Now add the condenser, just holding it onto the three-way adapter. Make a mental note of where the inner joint (end) of the condenser is (Fig. 18.6)— you'll want to put a clamp about there. Remove the condenser.

6. Get an extension clamp and clamp fastener. Open the jaws of the clamp so they're *wider* than the outer joint on a vacuum adapter. Set this clamp so that it can accept the outer joint of a vacuum adapter at the angle and height you made a mental note of in step 5.

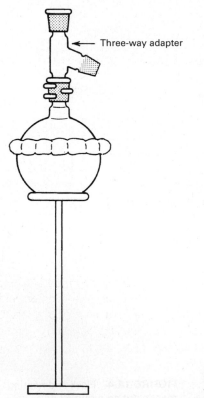

← Three-way adapter

FIGURE 18.5 Clamps and flask and three-way adapter.

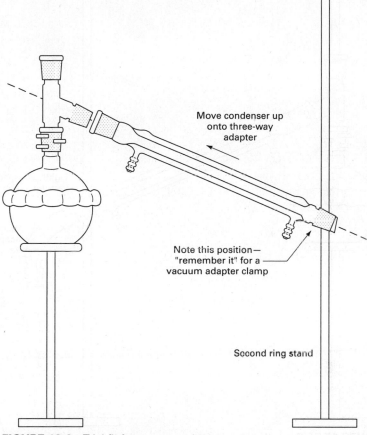

Move condenser up
onto three-way
adapter

Note this position—
"remember it" for a
vacuum adapter clamp

Second ring stand

FIGURE 18.6 Trial fit for vacuum adapter.

7. Put a vacuum adapter on the end of your condenser, and hold it there. Put the top of the condenser (outer joint) onto the three-way adapter, and get the vacuum adapter joint cradled in the clamp you've set for it. Push the condenser/vacuum adapter toward the three-way adapter; lightly tighten the clamp on the vacuum adapter. The clamp *must* stop the vacuum adapter from *slipping off the condenser* (Fig. 18.7).

8. At this point, you have two clamps, one holding the distilling flask and the other at the end of the condenser/vacuum adapter (Fig. 18.8).

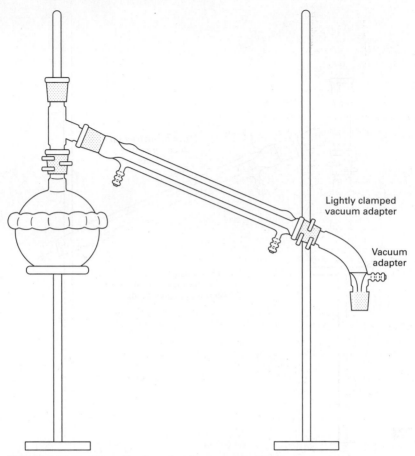

FIGURE 18.7 Correctly clamping the vacuum adapter.

9. There are two ways to go now:
   a. The receiving flask will not fit vertically from the end of the vacuum adapter to the bench top (Fig. 18.8). Easy. With the vacuum adapter clamp *relaxed* (not so loose that the adapter falls off!), rotate the adapter toward you. There's no reason it *has* to be at a right angle. Stick a receiving flask on the end, put a suitable cork ring under the flask, and that's it. Note that you can *easily* change receiving flasks. Just slide one off and slide another on.
   b. There's a lot of room from the end of the vacuum adapter to the bench top. In this case, you'll have to set up another clamp (possibly another ring stand) and clamp the flask at the neck (Fig. 18.9). This will make changing the receiving flask a bit more difficult; you'll have to unclamp one flask before slipping another in and reclamping the new one.

10. All the clamps are set up, all the joints are tight—now where is that thermometer adapter?

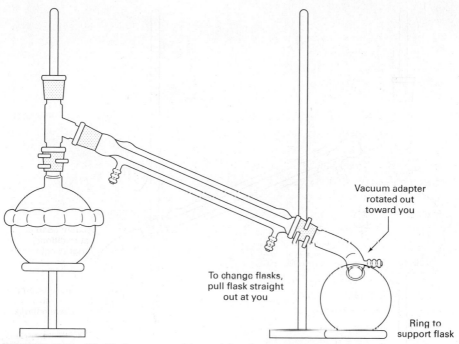

FIGURE 18.8   Distillation setup with receiving flask rotated toward you.

Labels in figure:
Vacuum adapter rotated out toward you
To change flasks, pull flask straight out at you
Ring to support flask

CLIPPING A DISTILLATION SETUP

There is another way of constructing a distillation setup: **Keck clips** (Fig. 18.10). These plastic clips have a large semicircle that gloms onto the outer joint, a smaller semicircle that hugs the inner joint, and three arms that hold the two semicircular clips and, as a consequence, hold the two parts of the glass joint together (Fig. 18.11). It's actually a lot of fun connecting jointware and clipping each joint together, watching the setup grow larger and larger as each new piece is clipped to the previous one.

As neat as Keck clips are, there is one small problem—they melt. If you clip together a distillation setup, and the compound in the flask boils at a relatively high temperature, the clip holding the distilling flask to the three-way adapter melts and then flows over the joint like melted cheese.

Never, ever use a plastic clip to attach a flask to a three-way adapter.

The clip that holds the three-way adapter to the condenser usually survives this, and the clips that hold the condenser to the vacuum adapter and the vacuum adapter to the receiving flask are not in any trouble at all.

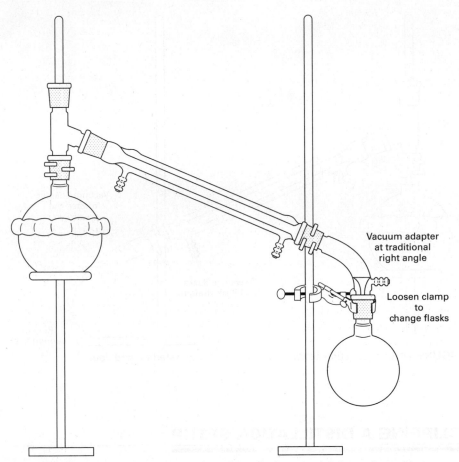

Vacuum adapter
at traditional
right angle

Loosen clamp
to
change flasks

FIGURE 18.9 Distillation setup with receiving flask at usual right-angle position.

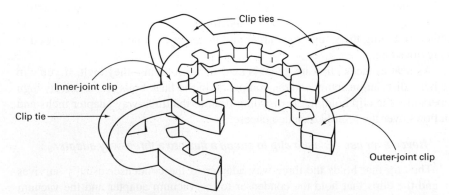

Clip ties

Inner-joint clip

Clip tie

Outer-joint clip

FIGURE 18.10 Top view and side view of a Keck clip.

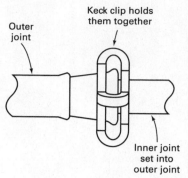

Keck clip holds
them together

Outer
joint

Inner joint
set into
outer joint   **FIGURE 18.11**   Hold on! It's the Keck clip in action.

The real fun starts when, after the setup cools and the plastic solidifies, you have to forcefully, yet carefully, chisel the globs of plastic off the joints.

Of course, this won't ever happen to you, oh reader of this manual, because you will use a clamp to hold the distilling flask, not a Keck clip. You know you may have to quickly lower the distilling flask, and if it's clipped, there's no way you can do that.

Chapter 19

DISTILLATION

- ■ *Gentle water flow through condensers—not Niagara Falls.*
- ■ *Water goes in the bottom, out the top.*
- ■ *Setups are three-dimensional: Flasks can rotate toward you, hoses upward, and so on.*

Separating or purifying liquids by vaporization and condensation is a very important step in one of our oldest professions. The word "still" lives on as a tribute to the importance of organic chemistry. There are two important points here:

1. *Vaporization.* Turning a liquid to a vapor.
2. *Condensation.* Turning a vapor to a liquid.

Remember these. They show up on quizzes.

But when do I use distillation? That is a very good question. Use the guidelines below to pick your special situation, and turn to that section. But you *should* read *all* of the sections anyway.

1. *Class 1: Simple distillation.* Separating liquids boiling below 150°C at one atmosphere (1 atm) from:
   a. Nonvolatile impurities.
   b. Another liquid that boils at least 25°C higher than the first. The liquids should dissolve in each other.
2. *Class 2: Vacuum distillation.* Separating liquids that boil above 150°C at 1 atm from:
   a. Nonvolatile impurities.
   b. Another liquid boiling at least 25°C higher than the first. The liquids should dissolve in each other.

3. **Class 3: Fractional distillation.** Separating liquid mixtures, soluble in each other, that boil at less than 25°C from each other at 1 atm.

4. **Class 4: Steam distillation.** Isolating tars, oils, and other liquid compounds that are *insoluble,* or slightly soluble, *in water at all temperatures.* Usually, natural products are steam distilled. They do *not* have to be liquids at room temperatures. (For example, caffeine, a solid, can be isolated from green tea.)

Remember, these are guides. If your compound boils at 150.0001°C, don't whine that you must do a vacuum distillation or both you and your product will die. I expect you to have some judgment and to pay attention to your instructor's specific directions.

DISTILLATION NOTES

1. Except for Class 4, steam distillation, two liquids that are to be separated must dissolve in each other. If they did not, they would form separable layers, which you could separate in a separatory funnel (see Chapter 15, "Extraction and Washing").

2. Impurities can be either **soluble** or **insoluble**. For example, the material that gives cheap wine its unique bouquet is soluble in the alcohol. If you distill cheap wine, you get clear grain alcohol separated from the "impurities," which are left behind in the distilling flask.

CLASS 1: SIMPLE DISTILLATION (FIG. 19.1)

For separation of liquids that boil below 150°C at 1 atm from:

1. Nonvolatile impurities.

2. Another liquid that boils at least 25°C higher than the first liquid. *The liquids must dissolve in each other.*

Sources of Heat

If one of the components boils below 70°C and you use a Bunsen burner, you may have a hard time putting out the fire. Use a steam bath or a heating mantle. Different distillations will require different handling (see Chapter 17, "Sources of Heat"). All distillations always require heating, so Chapter 17 is really closely tied to this section. This goes for enlightenment on using **boiling stones** and **clamps** as well (Chapter 18, "Clamps and Clamping").

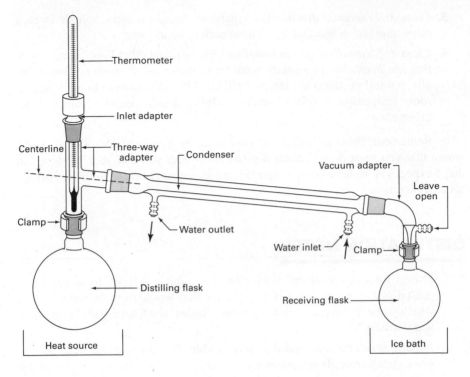

FIGURE 19.1 A complete, entire simple distillation setup.

The Three-Way Adapter

If there is any *one place* your setup will fall apart, here it is (Fig. 19.2). When you set up the jointware, it is important that all the joints *line up*. This is tricky, since, as you push one joint together, another pops right out. If you're not sure, call your instructor to inspect your work. Remember,

All joints must be tight!

The Distilling Flask

Choose a distilling flask carefully. If it's too big, you'll lose a lot of your product. If it's too small, you might have to distill in parts. *Don't fill the distilling flask more than half full.* Less than one-third full and you'll probably lose product. More than one-half full and you'll probably have undistilled material thrown up into the condenser (and into your previously clean product). Fill the distilling flask with the liquid you want to distill. You can remove the thermometer and thermometer adapter, fill the flask using a funnel, and then put the thermometer and its adapter back in place.

If you're doing a **fractional distillation** with a **column** (a Class 3 distillation), you should've filled the flask *before* clamping the setup. (Don't ever pour your mixture down a column. That'll contaminate everything!) You'll just have to

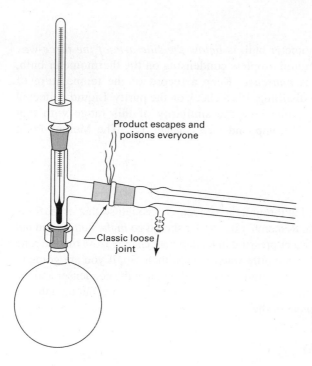

Product escapes and
poisons everyone

Classic loose
joint

FIGURE 19.2 The
"commonly-camouflaged-
until-it's-too-late" open joint.

disassemble some of the setup, fill the flask, reassemble what you've taken down, and pray that you haven't knocked all of the other joints out of line.

Put in a boiling stone if you haven't already. These porous little rocks promote bubbling and keep the liquid from superheating and flying out of the flask. This flying around is called **bumping**. Never drop a boiling stone into hot liquid, or you may be rewarded by having your body soaked in the hot liquid as it foams out at you.

Make sure all the joints in your setup are tight. Start the heat s-l-o-w-l-y until gentle boiling begins and liquid starts to drop into the receiving flask at the rate of about 10 drops per minute. *This is important.* If nothing comes over, you're not distilling, but merely wasting time. You may have to turn up the heat to keep material coming over.

The Thermometer Adapter

Read all about it. Ways of having fun with thermometer adapters have been detailed (see text accompanying Figs. 4.7–4.11 in Chapter 4, "Jointware").

The Ubiquitous Clamp

A word about clamps. *Use!* They can save you $68.25 in busted glassware (see Chapter 18, "Clamps and Clamping").

The Thermometer

Make sure the entire thermometer bulb is *below the side arm of the three-way adapter.* If you don't have liquid droplets condensing on the thermometer bulb, the temperature you read is *nonsense.* Keep a record of the temperature of the liquid or liquids that are distilling. It's a check on the purity. Liquid collected over a 2°C range is fairly pure. Note the similarity of this range with that of the *melting point* of a pure compound (see Chapter 12, "The Melting-Point Experiment").

The Condenser

Always keep cold water running through the condenser, enough so that *at least the lower half is cold* to the touch. Remember that water should go *in the bottom* and *out the top* (Fig. 19.1). Also, the water pressure in the lab may change from time to time and usually goes up at night, since little water is used then. So, if you are going to let condenser cooling water run overnight, tie the tubing on at the condenser and the water faucet with wire or something. And if you don't want to flood out the lab, see that the outlet hose can't flop out of the sink.

The Vacuum Adapter

It is important that the tubing connector remain *open to the air;* otherwise, the entire apparatus will, quite simply, explode.

> *Warning: Do not just stick the vacuum adapter on the end of the condenser and hope that it will not fall off and break.*

This is foolish. I have no sympathy for anyone who will not use clamps to save their own breakage fee. They deserve to pay.

The Receiving Flask

The receiving flask should be large enough to collect what you want. You may need several, and they may have to be changed during the distillation. Standard practice is to have one flask ready for what you are going to throw away and others ready to save the stuff you want to save.

The Ice Bath

Why everyone insists on loading up a bucket with ice and trying to force a flask into this mess, I'll never know. How much cooling do you think you're going to get with just a few small areas of the flask barely touching ice? Get a suitable receptacle—a large beaker, enameled pan, or whatever. *It should not leak.* Put it under the flask. Put some water in it. *Now add ice.* Stir. Serves four.

Ice bath really means ice-water bath.

THE DISTILLATION EXAMPLE

Say you place 50 mL of liquid A (bp 50°C) and 50 mL of liquid B (bp 100°C) in a 250-mL round-bottom flask. You drop in a boiling stone, fit the flask in a distillation setup, and turn on the heat. Bubbling starts, and soon droplets form on the thermometer bulb. The temperature shoots up *from room temperature to about 35°C,* and a liquid condenses and drips into the receiver. That's bad. The temperature should be close to 50°C. This low-boiling material is the **forerun** of a distillation, and you won't want to keep it.

Keep letting liquid come over until the temperature stabilizes at about 49°C. Quick! Change receiving flasks now!

The new receiving flask is on the vacuum adapter, and the temperature is about 49°C. Good. Liquid comes over, and you heat to get a rate of about 10 drops per minute collected in the receiver. As you distill, the temperature slowly increases to maybe 51°C and then starts moving up rapidly.

Here you stop the distillation and change the receiver. Now in one receiver you have a *pure liquid, bp 49–51°C.* Note this **boiling range**. It is just as good a test of purity as a melting point is for solids (see Chapter 12, "The Melting-Point Experiment").

Always report a boiling point for liquids as routinely as you report melting points for solids. The boiling point is actually a **boiling range** and should be reported as such:

"bp 49–51°C"

Now, if you put on a new receiver and start heating again, you may discover *more material coming over at 50°C!* Find that strange? Not so. All it means is that you were distilling too rapidly and some of the low-boiling material was left behind. It is very difficult to avoid this situation. Sometimes it is best to ignore it, unless a yield is very important. You can combine this "new" 50°C fraction with the other good fraction.

For liquid B, boiling at 100°C, merely substitute some different boiling points and go over the same story.

THE DISTILLATION MISTAKE

OK, you set all this stuff up to do a distillation. Everything's going fine. Clamps in the right place. No arthritic joints, even the vacuum adapter is clamped on, and the thermometer is at the right height. There's a bright golden haze on the meadow, and everything's going your way. So, you begin to boil the liquid. You even remembered the boiling stone. Boiling starts slowly, then more rapidly. You think, "This is *it!*" Read that temperature, now. Into the notebook, "The mixture started boiling at 26°C."

And you are dead wrong.

What happened? Just ask—

Is there liquid condensing on the thermometer bulb??
No!

So, congratulations, you've just recorded the room temperature. There are days when more than half of the class will report distillation temperatures as "Hey, I see it starting to boil now" temperatures. Don't participate. Just keep watching as the liquid boils. Soon, droplets *will* condense on the thermometer bulb. The temperature will go up quickly, and then *stabilize.* Now read the temperature. That's the boiling point. But wait! It's *not* a distillation temperature until that first drop of liquid falls into the receiving flask.

Just before you leave for your next class, **vacuum distillation**, you might want to check out the online chapter "Theory of Distillation" on the Book Companion Site at www.wiley.com/college/zubrick.

CLASS 2: VACUUM DISTILLATION

For separation of liquids that boil above 150°C at 1 atm from:

1. Nonvolatile impurities.

2. Another liquid that boils at least 25°C higher than the first liquid. The liquids must dissolve in each other. This is like the **simple distillation** with the changes shown (Fig. 19.3).

Why vacuum distill? If the substances boil at high temperatures at 1 atm, they may decompose when heated. Putting the liquid under vacuum makes the liquid boil at a lower temperature. With the pressure reduced, there are fewer molecules in the

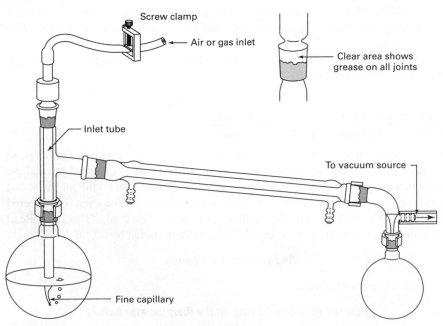

FIGURE 19.3 A vacuum distillation setup.

way of the liquid you are distilling. Since the molecules require less energy to leave the surface of the liquid, *you can distill at a lower temperature,* and *your compound doesn't decompose.*

Pressure Measurement

If you want to measure the pressure in your vacuum distillation setup, you'll need a **closed-end manometer**. There are a few different types, but they all work essentially in the same way. I've chosen a "stick" type (Fig. 19.4). This particular model

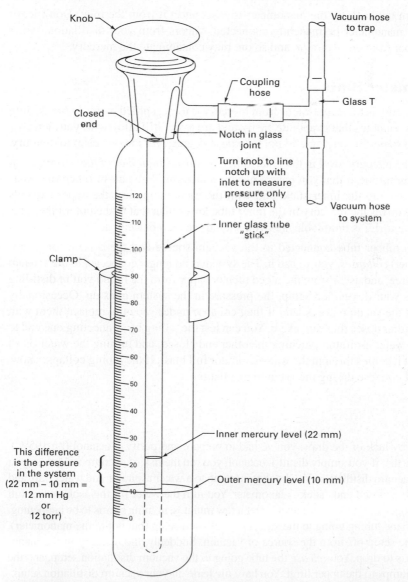

FIGURE 19.4   A closed-end "stick" manometer.

needs help from a short length of rubber tubing and a glass T to connect it to the vacuum distillation setup.

1. Turn on the source of vacuum and wait a bit for the system to stabilize.
2. Turn the knob on the manometer so that the notch in the joint lines up with the inlet.
3. Wait for the mercury in the manometer to stop falling.
4. Read the *difference* between the inner and outer levels of mercury. This is the system pressure, literally in **millimeters of mercury**, which we now call *torr*.
5. Turn the knob on the manometer to disconnect it from the inlet. Don't leave the manometer permanently connected. Vapors from your distillation, water vapor from the aspirator, and so on, may contaminate the mercury.

Manometer Hints

1. Mercury is toxic, the vapor from mercury is toxic, spilled mercury breaks into tiny globules that evaporate easily and are toxic, it'll alloy with your jewelry, and so on. Be very careful not to expose yourself (or anyone else) to mercury.
2. If the mercury level in the *inner* tube goes *lower than that of the outer tube,* it does not mean that you have a negative vacuum. Some air or other vapor has gotten into the inner stick, and with the vacuum applied, the vapor expands and drives the mercury in the inner tube lower than that in the outer tube. This manometer is unreliable, and you should seek a replacement.
3. If a rubber tube connected to the vacuum source and the system (or manometer) *collapses,* you've had it. The system is no longer connected to the vacuum source, and as air from the bleed tube or vapor from the liquid you're distilling fills your distillation setup, the pressure in the system goes up. Occasionally test the vacuum hoses, and if they collapse under vacuum, replace them with sturdier hoses that can take it. You can test the tubing by connecting one end to the water aspirator, pinching the other end closed, and turning the water on so that it comes through the water aspirator full blast. If the tubing collapses now, it'll collapse during the vacuum distillation.

Leaks

Suppose, by luck of the draw, you've had to prepare and purify 1-octanol (bp 195°C). You know that if you simply distill 1-octanol, you run the risk of decomposing it, so you set up a vacuum distillation. You hook your setup to a water aspirator and water trap, and you attach a closed-end "stick" manometer. You turn the water for the aspirator on full blast and open the stick manometer. After a few minutes, nothing seems to be happening. You *pinch* the tubing going to the vacuum distillation setup (but not to the manometer), closing the setup off from the source of vacuum. Suddenly, the mercury in the manometer starts to drop. You *release* the tube going to the vacuum distillation setup, and the mercury jumps to the upper limit. You have **air leaks** in your vacuum distillation setup.

Air leaks can be difficult to find. At best, you push some of the joints together again and the system seals itself. At worst, you have to take apart all the joints and regrease every one. Sometimes you've forgotten to grease all the joints. Often a joint has been etched to the point that it cannot seal under vacuum, though it is perfectly fine for other applications. Please get help from your instructor.

Pressure and Temperature Corrections

You've found all the leaks, and the pressure in your vacuum distillation setup is, say, 25 torr. Now you need to know the boiling point of your compound, 1-octanol, this time at 25 torr, not 760 torr. You realize it'll boil at a lower temperature, but just how low? The handy nomographs can help you estimate the new boiling point.

This time you have the boiling point at 760 torr (195°C) and the pressure you are working at (25 torr). Using Fig. 19.5, you:

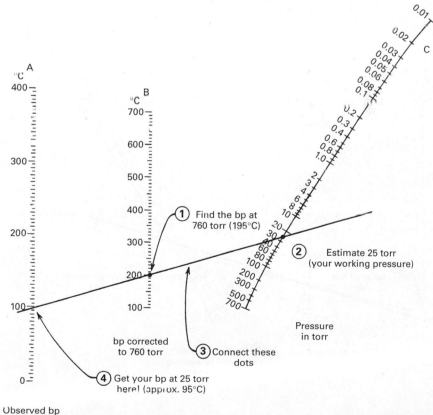

FIGURE 19.5   One-point conversion.

1. Find the boiling point at 760 torr (195°C) on line B (the middle one).

2. Find the pressure you'll be working at (25 torr) on line C (the one on the far right). You'll have to estimate this point.

3. Line up these two points using a straightedge and see where the straightedge cuts the observed boiling point line (line A, far left). I get about 95°C.

So, a liquid that boils at 195°C at 760 torr will boil at about 95°C at 25 torr. Remember, this is an estimate.

Now suppose you looked up the boiling point of 1-octanol and all you found was: 98[19]. This means that the boiling point of 1-octanol is 98°C at 19 torr. Two things should strike you.

1. This is a *higher* boiling point at a *lower* pressure than we've gotten from the nomograph.

2. I wasn't kidding about this process being an estimation of the boiling point.

Now we have a case of having an observed boiling point at a pressure that is *not* 760 torr (1-octanol again; 98°C at 19 torr). We'd like to get to 25 torr, our working pressure. This requires a double conversion, as shown in Fig. 19.6.

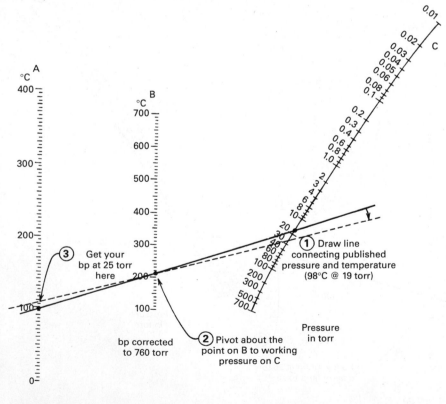

Observed bp

FIGURE 19.6 Two-point conversion.

1. On the observed bp line (line A), find 98°C.
2. On the pressure-in-torr line (line C), find 19.
3. Using a straightedge, connect those points. Read the bp corrected to 760 torr (line B): I get 210°C.
4. Now, using the 210°C point as a fulcrum, pivot the straightedge until the 210°C point on line B and the pressure you're working at (25 torr) on line C line up. You see, you're in the same position as in the previous example with a "corrected to 760 torr bp" and a working pressure.
5. See where the straightedge cuts the observed boiling point line (line A). I get 105°C.

So, we've estimated the boiling point at about 105°C at 25 torr. The last time it was 95°C at 25 torr. Which is it? Better you should say you *expect* your compound to come over at 95–105°C. Again, this is not an unreasonable expectation for a vacuum distillation.

The pressure–temperature nomograph (Fig. 19.7) is really just a simple, graphical application of the Clausius–Clapeyron equation. If you know the heat of vaporization of a substance, as well as its normal boiling point, you can calculate the boiling

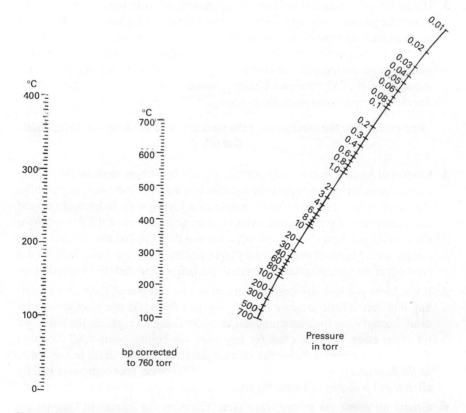

FIGURE 19.7   A clean nomograph for your own use.

point at another temperature. You do have to assume that the heat of vaporization is constant over the temperature range you're working with, and that's not always so. Where's the heat of vaporization in the nomograph? One is built into the slopes and spacings on the paper. And, yes, that means that the heat of vaporization is forced to be the same for all compounds, be they alkanes, aldehydes, or ethers. So do not be surprised at the inaccuracies in this nomograph; be amazed that it works as well as it does.

Vacuum Distillation Notes

1. Read all of the notes on Class 1 (simple distillation).

2. The thermometer can be replaced by a **gas inlet tube**. It has a long, fine capillary at one end (Fig. 19.3). This helps to stop the *extremely bad bumping* that goes along with vacuum distillations. The fine stream of bubbles through the liquid produces the same results as a boiling stone. Boiling stones are useless, since all the adsorbed air is whisked away by the vacuum and the nucleating cavities plug up with liquid. The fine capillary does not let in a lot of air, so we are doing a vacuum distillation anyway. Would you be happier if I called it a **reduced-pressure** distillation? An inert gas (nitrogen?) may be let in if the compounds decompose in air.

3. If you can get a **magnetic stirrer** and **magnetic stirring bar**, you won't have to use the gas inlet tube approach. Put a magnetic stirring bar in the flask with the material you want to vacuum distill. Use a *heating mantle* to heat the flask, and put the magnetic stirrer under the mantle. When you turn the stirrer on, a magnet in the stirrer spins, and the stirring bar (a Teflon-coated magnet) spins. Admittedly, stirring through a heating mantle is not easy, but it can be done. *Stirring the liquid also stops the bumping.*

   Remember, first the stirring, then the vacuum, then the heat—or WHOOSH! Got it?

4. Control of heating is *extremely critical.* I don't know how to shout this loudly enough on paper. Always apply the vacuum first and watch the setup for a while. Air dissolved or trapped in your sample or a highly volatile leftover (maybe ethyl ether from a previous extraction) can come flying out of the flask *without the heat.* If you heated such a setup a bit and then applied the vacuum, your sample would blow all over, possibly right into the receiving flask. Wait for the contents of the distilling flask to calm down before you start the distillation.

5. If you know you have low-boiling material in your compound, think about distilling it at atmospheric pressure first. If, say, half the liquid you want to vacuum distill is ethyl ether from an extraction, consider doing a simple distillation to get rid of the ether. Then the ether (or any other low-boiling compound) won't be around to cause trouble during the vacuum distillation. If you distill first at 1 atm, *let the flask cool before you apply the vacuum.* Otherwise, your compound will fly all over and probably will wind up, undistilled and impure, in your receiving flask.

6. *Grease all joints, no matter what* (see "Greasing the Joints" in Chapter 4). Under vacuum, it is easy for any material to work its way into the joints and turn into concrete, and the joints will never, ever come apart again.

7. The **vacuum adapter** is connected to a **vacuum source**, either a **vacuum pump** or a **water aspirator**. Real live vacuum pumps are expensive and rare and not usually found in the undergraduate organic laboratory. If you can get to use one, that's excellent. See your instructor for the details. The **water aspirator** is used lots, so read up on it (see Chapter 13).

8. During a vacuum distillation, it is not unusual to collect a *pure compound over a 10–20°C temperature range*. If you don't believe it, you haven't ever done a vacuum distillation. It has to do with pressure changes throughout the distillation because the setup is far from perfect. Although a vacuum distillation is not difficult, it requires peace of mind, large quantities of patience, and a soundproof room to scream in so you won't disturb others.

9. A **Claisen adapter** in the distilling flask lets you take temperature readings and can help stop your compound from splashing over into the distillation receiver (Fig. 19.8). On a microscale, though, the extra length of the Claisen adapter could hold up a lot of your product. Check with your instructor. You could also use a **three-neck flask** (Fig. 19.9). Think! And, of course, use some glassware too.

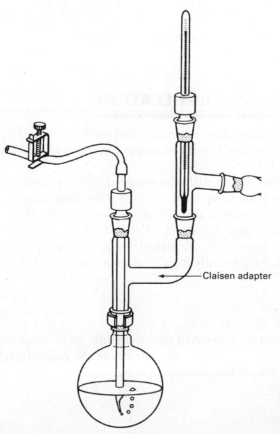

—Claisen adapter

FIGURE 19.8 A Claisen adapter. You can vacuum distill and take temperatures, too.

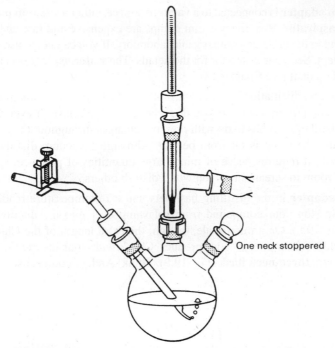

One neck stoppered

FIGURE 19.9 Multipurpose setup with a three-neck flask.

CLASS 3: FRACTIONAL DISTILLATION

For separation of liquids that are soluble in each other and boil fewer than 25°C from each other, use fractional distillation. This is like simple distillation with the changes shown (Figs. 19.10 and 19.11).

Fractional distillation is used when the components to be separated boil *within 25°C of each other.* Each component is called a **fraction.** Clever where they get the name, eh? This temperature difference is not gospel. And don't expect terrific separations, either. Let's just leave it at *close boiling points.* How close? That's hard to answer. "Is an orange?" That's easier to answer. If the experiment tells you to "fractionally distill," at least you'll be able to set it up right.

How This Works

If one distillation is good, two are better. And fifty, better still. So you have lots and lots of little, tiny distillations occurring on the surfaces of the **column packing**, which can be glass beads, glass helices, ceramic pieces, metal chips, or even stainless steel wool.

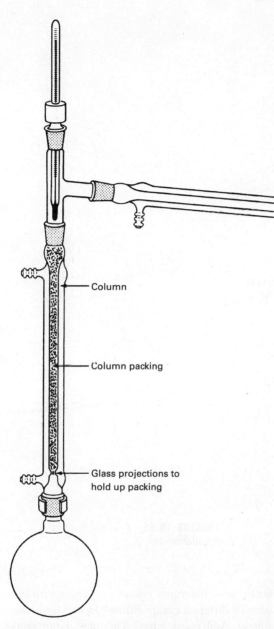

- Column
- Column packing
- Glass projections to hold up packing

FIGURE 19.10   The fractional distillation setup.

As you heat your mixture, it boils, and the vapor that comes off this liquid is *richer in the lower-boiling component.* The vapor moves out of the flask and condenses, say, on the first centimeter of column packing. Now, the composition of the liquid still in the flask has changed a bit—it is *richer in the higher-boiling*

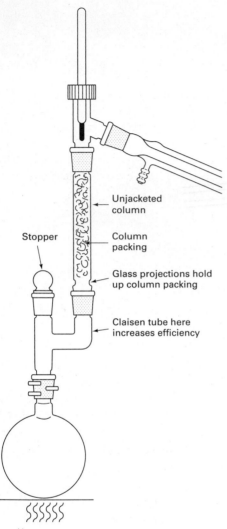

Unjacketed
column

Stopper

Column
packing

Glass projections hold
up column packing

Claisen tube here
increases efficiency

Heat source

FIGURE 19.11 Another fractional
distillation setup.

component. As more of *this* liquid boils, more hot vapor comes up, mixes with the
first fraction, and produces a new vapor of different composition—*richer yet in the
more volatile (lower-boiling) component.* And guess what? This new vapor condenses in the *second centimeter* of column packing. And again, and again, and again.

Now all of these are **equilibrium steps**. It takes some time for the fractions to move up the column, get comfortable with their surroundings, meet the
neighbors. . . . And if you *never* let any of the liquid–vapor mixture out of the
column, a condition called **total reflux**, you might get a single pure component
at the top—namely, the lower-boiling, more volatile component all by itself! This
is an ideal separation.

Fat lot of good that does you when you have to hand in a sample. So, you turn up the heat, let some of the vapor condense, and *take off this top fraction.* This raises hell in the column. *Nonequilibrium conditions abound—mixing. Arrrgh!* No more completely pure compound. And the faster you distill, the faster you let material come over, the higher your **throughput**—the worse this gets. Soon you're at **total takeoff**, and there is no time for an equilibrium to become established. And if you're doing that, you shouldn't even bother using a column.

You must strike a compromise. Fractionally distill *as slowly as you can,* keeping in mind that eventually the lab does end. Slow down your fractional distillations; I've found that 5–10 drops per minute coming over into the receiving flask is usually suggested. It will take a bit of practice before you can judge the best rate for the best separation. See your instructor for advice.

Fractional Distillation Notes

1. Read all of the notes on Class 1 (simple distillation).

2. Make *sure* you have not confused the **column** with the **condenser**. The *column is wider and has glass projections inside* at the bottom to hold up the packing.

3. *Don't break off the projections!*

4. Do not run water through the jacket of the column!

5. Sometimes, the column is used *without* the column packing. This is all right, too.

6. If it is necessary, and it usually is, push a wad of heavy metal wool down the column, *close to the support projections,* to support the packing chips. Sometimes the packing is entirely this stainless steel wool. You can see that it is self-supporting.

7. Add the column packing. Shake the column lightly to make sure none of the packing will fall out later into your distillation flask.

8. With all the surface area of the packing, a lot of liquid is *held up* on it. This phenomenon is called **column holdup**, *since it refers to the material retained in the column. Make sure you have enough compound to start with, or it will all be lost on the packing.*

9. A **chaser solvent** or **pusher solvent** is sometimes used to help blast your compound off the surface of the packing material. It should have a *tremendously high boiling point relative to what you were fractionating.* After you've collected most of one fraction, some of the material is left on the column. So, you throw this chaser solvent into the distillation flask, fire it up, and start to distill the chaser solvent. As the chaser solvent comes up the column, it heats the packing material, your compound is blasted off the column packing, and more of your compound comes over. Stop collecting when the temperature starts to rise—that's the chaser solvent coming over now. As an example, you might expect *p*-xylene (bp 138.4°C) to be a really good chaser, or pusher, for compounds that boil at less than, say, 100°C.

But you have to watch out for the deadly azeotropes.

AZEOTROPES

Once in a while, you throw together two liquids and find that you cannot separate part of them. And I don't mean because of poor equipment, or poor technique, or other poor excuses. You may have an **azeotrope**, a mixture with a *constant boiling point.*

One of the best-known examples is ethyl alcohol–water. This 95.6% ethanol–4.4% water solution will boil to dryness, at a *constant temperature.* It's slightly scary, since you learn that a liquid is a pure compound if it boils at a constant temperature. And you thought you had it made.

There are two types of azeotropes. If the azeotrope boils off first, it's a **minimum-boiling azeotrope**. After it's all gone, if there is any other component left, only then will that component distill.

If any of the components come off first, and then the azeotrope, you have a **maximum-boiling azeotrope**.

Quiz question:

Fifty milliliters of a liquid boils at 74.8°C from the beginning of the distillation to the end. Since there is no wide boiling range, can we assume that the liquid is pure?

No. It may be a constant-boiling mixture called an azeotrope.

You should be able to see that you have to be really careful in selecting those chaser or pusher solvents mentioned. Sure, water (bp 100°C) is hot enough to chase ethyl alcohol (bp 78.3°C) from any column packing. Unfortunately, water and ethyl alcohol form an azeotrope and the technique won't work. (*Please see the online chapter "Theory of Distillation" on the Book Companion Site at www.wiley.com/college/zubrick.*)

CLASS 4: STEAM DISTILLATION

Mixtures of tars and oils must not dissolve well in water (well, not much, anyway), so we can steam distill them. The process is pretty close to simple distillation, but you should have a way of getting *fresh hot water into the setup* without stopping the distillation.

Why steam distill? If the stuff you're going to distill is *only slightly soluble in water* and may decompose at its boiling point and the bumping will be terrible with a vacuum distillation, it is better to **steam distill**. Heating the compound in the presence of steam makes the compound boil at a lower temperature. This has to do with partial pressures of water and organic oils and such.

There are two ways of generating steam: externally and internally.

External Steam Distillation

In an **external steam distillation**, you lead steam from a steam line through a water trap, and thus into the system. The steam usually comes from a **steam tap** on the bench top. This is classic. This is complicated. This is dangerous.

1. Set up your external steam distillation apparatus in its entirety. Have *every-thing* ready to go. This includes having the material you want to distill in the distilling flask, the steam trap already attached, condensers up and ready, a

large receiving flask, and so on. All you should have to do is attach a single hose from the steam tap to your steam trap and start the steam.

2. *Have your instructor check your setup before you start!* I cannot shout this loudly enough on this sheet of paper. Interrupting an external steam distillation, just because you forgot your head this morning, is a real trial.

3. Connect a length of rubber tubing to your bench steam outlet, and lead the rubber tubing into a drain.

4. Now, watch out! Slowly, carefully, open the steam stopcock. Often you'll hear clanging, bonking, and thumping, and a mixture of rust, oil, and dirtladen water will come spitting out. Then some steam bursts come out, and, finally, you have a stream of steam. Congratulations. You have just *bled the steam line.* Now close the steam stopcock, wait for the rubber tubing to cool a bit, and then. . . .

5. Carefully (**Caution**—may be hot!) attach the rubber tubing from the steam stopcock to the inlet of your steam trap.

6. Open the steam trap drain, and then carefully reopen the bench steam stopcock. Let any water drain out of the trap, then carefully close the drain clamp. Be careful.

7. You now have steam going through your distillation setup, and as soon as product starts to come over, you'll be doing an external steam distillation. Periodically, open the steam trap drain (**Caution**—hot!), and let the condensed steam out.

8. Apparently you can distill as fast as you can let the steam into your setup, *as long as all of the steam condenses* and doesn't go out into the room. Sometimes when you steam distill, you need to hook two condensers together, making a very long supercondenser. Check with your instructor.

9. When you're finished (see "Steam Distillation Notes" following), turn off the steam, let the apparatus cool, and dismantle everything.

You can use many types of steam traps with your distillation setup. I've shown two (Figs. 19.12 and 19.13), but these are not the only ones, and you may use something different. The point is to note the **steam inlet** and the **trap drain** and how to use them.

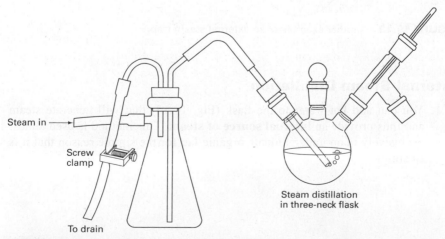

Steam in →

Screw clamp

Steam distillation in three-neck flask

To drain

FIGURE 19.12   One example of an external steam trap.

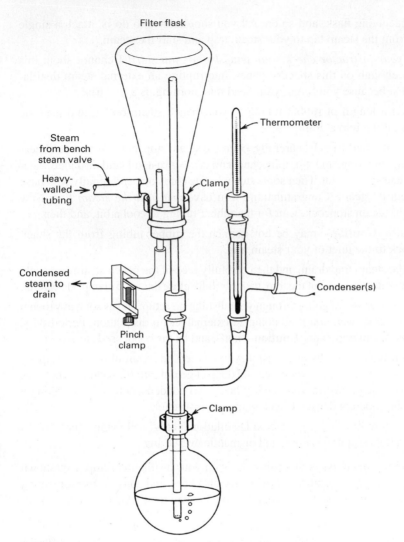

Filter flask

Thermometer

Steam
from bench
steam valve

Heavy-
walled
tubing

Clamp

Condensed
steam to
drain

Condenser(s)

Pinch
clamp

Clamp

FIGURE 19.13 Another example of an external steam trap.

Internal Steam Distillation

1. You can add hot water to the flask (Fig. 19.14) that will generate steam and thus provide an **internal source of steam**. This method is used almost exclusively in an undergraduate organic lab for the simple reason that it is so simple.

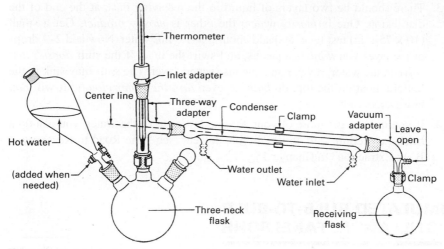

FIGURE 19.14   An internal steam distillation setup.

2. Add to the distilling flask at least three times as much water (maybe more) as sample. Do not fill the flask much more than half full (three quarters, maybe). You've got to be careful. Very careful.

3. Periodically, add more *hot water* as needed. When the water boils and turns to steam, it also leaves the flask, carrying product.

STEAM DISTILLATION NOTES

1. Read all of the notes on Class 1 distillations.

2. Collect some of the distillate, the stuff that comes over, in a small test tube. Examine the sample. If you see *two layers,* or *if the solution is cloudy,* you're not done. Your product is still coming over. Keep distilling and keep adding hot water to generate more steam. *If you don't see any layers, don't assume you're done.* If the sample is slightly soluble in the water, the two layers or cloudiness might not show up. Try *salting out.* This has been mentioned before in connection with **extraction** and **recrystallization** as well (see "Salting Out" in Chapter 13 and "Extraction Hints" in Chapter 15). Add some salt to the solution you've collected in the test tube, shake the tube to dissolve the salt, and, if you're lucky, more of your product may be squeezed out of the aqueous layer, forming a *separate layer.* If that happens, *keep steam distilling* until the product does not come out when you treat a test solution with salt.

3. There should be two layers of liquid in the receiving flask at the end of the distillation. One is *mostly water;* the other is *mostly product.* Get a small (10 × 75-mm) test tube, and add about 0.5–1.0 mL water. Now add 2–4 drops of the layer you *think* is aqueous, and swirl the tube. If the stuff *doesn't dissolve* in the water, it's *not* an aqueous (water) layer. The stuff may sink to the bottom, float on the top, do *both,* or even *turn the water cloudy!* It will *not,* however, dissolve.

4. If you have to get more of your organic layer out of the water, you can do a **back-extraction** with an immiscible solvent (see "The Road to Recovery— Back-Extraction" in Chapter 15).

SIMULATED BULB-TO-BULB DISTILLATION: FAKELROHR

If you have to distill small amounts and you don't want to lose a lot of product on the surface of condensers, adapters, and all that, you might use a Kugelrohr apparatus (Fig. 19.15). The crude material, in a flask in an oven, distills from bulb-to-bulb with very little loss. As a bonus, the entire stack of bulbs (and flask) rotate à la rotary evaporator (see "The Rotary Evaporator," Chapter 21), so the material you're distilling coats the flask and distills off the much larger surface of the flask walls. Nice, but expensive, and with a lab full of people, quite a holdup if all you have is one unit.

What to do? Well, we just have to fake it. You connect two flasks by a vacuum adapter and distilling head, and voilà: Fakelrohr. OK, it's not true bulb-to-bulb distillation, and they don't rotate, but it's the shortest-path distillation we can set up in the undergraduate organic laboratory outside of the microscale Hickman still (see "Microscale Distillation," Chapter 20). Fakelrohr setups for jointware, microscale jointware, and Williamson test tubeware are shown in Fig. 19.16.

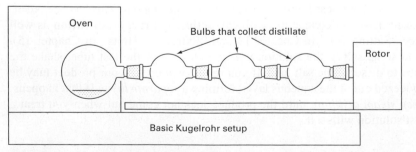

FIGURE 19.15 Simple schematic of a Kugelrohr bulb-to-bulb distillation apparatus.

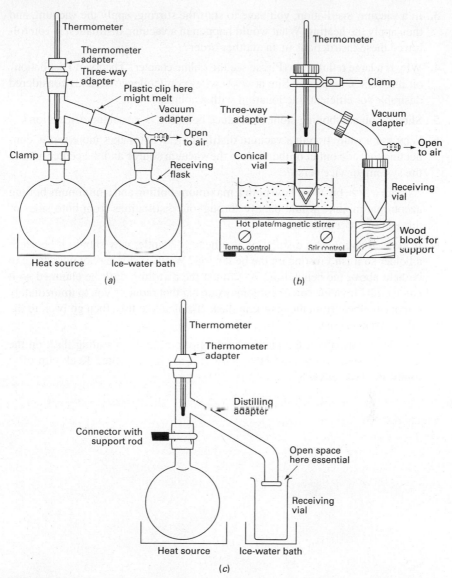

FIGURE 19.16 Fakelrohr setups: (*a*) jointware, (*b*) microscale, and (*c*) Williamson test tube.

EXERCISES

1. Often a clear product, upon distillation, initially gives a cloudy liquid. This observation has been called a "reverse steam distillation." Explain.

2. Define boiling point.

3. In a vacuum distillation, you have to start the stirring, apply the vacuum, and then apply the heating. What would happen in a vacuum distillation if you followed these instructions in an another order?

4. Why is a large reflux ratio (Please see the online chapter "Theory of Distillation" on the Book Companion Site at www.wiley.com/college/zubrick.) considered desirable for efficient fractionation with a column?

5. Should you add boiling stones to reduce bumping in a vacuum distillation?

6. Why use a water trap in a vacuum distillation? What makes more sense, connecting the side outlet of the flask to the vacuum source and the center tube to the system, or vice versa?

7. Explain a little bit about how both a maximum-boiling and a minimum-boiling azeotropic mixture would behave throughout distillations from both sides of the azeotropic temperature.

8. If the heat source in a distillation is a heating mantle, why should it be set in a holder so it isn't resting on the bench top? Would a stirring hot plate have to be held above the bench top? Why must the distilling flask be clamped so it cannot fall? Imagine some problem occurring that requires you to immediately remove the heat from the distilling flask. Think about this, then go back to the other two questions.

9. You don't normally use a plastic Keck clip to hold the distilling flask on the rest of the setup. Why not? How hard is it to chip a melted Keck clip off a jointware flask, anyway?

MICROSCALE DISTILLATION

LIKE THE BIG GUY

Before you read this chapter, read up on "Class 1: Simple Distillation" (Chapter 19), so you'll know what I'm talking about, so you can set up the same apparatus and do the same distillation for the same classes you've always done.

That's not completely true. Using this apparatus at the true microscale (0.01 mole), you have two drops of a liquid to distill. I'd see my instructor before I did any of these techniques with fewer than 10 mL.

Class 1: Simple Distillation

Perhaps you'll be using a conical vial rather than a round-bottom flask, but the setup and operation are the same as with the larger apparatus. *Note:* The tubing carrying cooling water can be so unwieldy that moving the tubing can pull this tiny setup out of alignment.

Class 2: Vacuum Distillation

Because the setup is so small and your sample size is so small, you've got to be very careful with vacuum distillation. With ℭ 14/20 ware (it's a little bigger), the operations are identical to those with the ℭ 24/40 or ℭ 19/22 sizes. I don't *have* the ℭ 14/10 setup yet, so I can only extrapolate. You'll have the same entertaining experiences. Watch the vacuum hose, too. It can be so unwieldy that it pulls the apparatus out of line if it's moved.

Oh: Check out "The O-Ring Cap Seal" discussion for a perspective on microscale vacuum distillation (see Chapter 5, "Microscale Jointware").

Class 3: Fractional Distillation

You'll probably substitute one of those **air condensers** for the distilling column. As a straight, unjacketed tube, the air condenser is not very efficient. I'd be careful about packing it with anything; the column holdup, as small as it is, is likely to hold up a lot of your product.

Class 4: Steam Distillation

With sample sizes this small, internal steam distillation is the only way out. With the microscale equipment, you can inject water by syringe rather than opening up the apparatus to pour water in, and lose less product. Again, if your sample is very small, see your instructor about any problems.

MICROSCALE DISTILLATION II: THE HICKMAN STILL

If you have only 1–2 mL to distill, then the Hickman still is probably your only way out (Fig. 20.1).

The Hickman Still Setup

1. Set up the stirring hot plate and sand bath. I suppose you *could* stick a thermometer in the sand to give you an idea of how warm the sand is.

2. Put the sample you want to distill into an appropriately sized conical vial. Two things: Don't fill the vial more than halfway, and don't use such a big vial that the product disappears. Less than one-third full, and you'll lose a lot on the walls of the vial.

3. These vials can fall over easily. Keep them in a holder, or clamp them until you're ready to use them.

4. Add a microboiling stone *or* a spin vane, not both, to the vial.

5. Get a clamp ready to accept the Hickman still at about the position you want, just open and loosely clamped—horizontally—to the ring stand.

6. Scrunch the conical vial down into the sand, and push it to where you want it to be. Hold the vial at the top.

7. Put the Hickman still on the vial. Now, even though you have a sand buttress, this baby is still top-heavy and can fall. Hold the vial *and* the still somewhere about where they meet.

8. Jockey the clamp and/or the vial and still so that they meet. Clamp the still at the top in the clamp jaws. Get the vial and still in their final resting place; tighten the clamp to the ring stand.

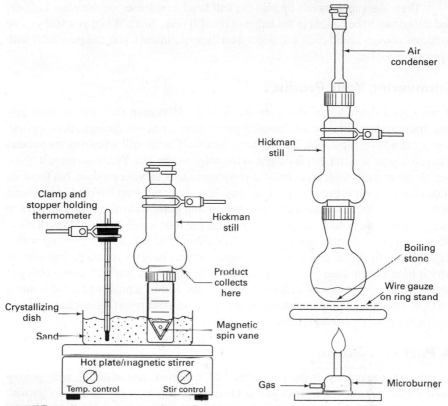

FIGURE 20.1   Two different Hickman still setups.

9. Hang a thermometer by a clamp and stopper so that the thermometer goes straight down the throat of the still. Place the thermometer bulb at about the joint of the vial and still. Do *not* let the thermometer touch the sides of the vial or still.

10. You're ready to heat.

Hickman Still Heating

You don't have very much liquid here, so it doesn't take much heat to boil your sample. If you have no idea what temperature your sample will boil at, you should just heat carefully until you see it boil, and then *immediately* turn the heat down 10–20%. Of course, you have a microboiling stone or magnetic spin vane in there, don't you?

As you heat the sample, you'll see a ring of condensate travel up the vial, and when it hits the thermometer bulb, the temperature reading should move up. As usual, if there's **refluxing liquid** on the bulb and the temperature has stabilized, that's the boiling point.

Then, the vapor travels up into the still head, condenses on the glass surface, and falls down to be caught in the bulge of the still (Fig. 20.1). When you feel you've collected enough liquid (this will depend on the experiment), you can go after it with a pipet and rubber bulb.

Recovering Your Product

If you have a thermometer down the throat of the **Hickman still**, forget about getting your product out with a disposable pipet; there's just not enough room, period. If you use a $5\frac{3}{4}$-in. pipet, you'll likely snap the tip off in the still when you try to bend it to reach your product that is caught in the bulge of the still. Yes, you *can* just about wangle the end of a 9-in. pipet into the crevice and pick up your product, but there's a good chance of breaking the tip off anyway. You might wind up having to put a bend in the tip by using a burner and your considerable glassworking skills. You have to use a small burner and a small flame and heat the tube carefully so it doesn't close off or melt off—in short, see your instructor about this. You should wind up with a slightly bent tip at the end of the 9-in. pipet. What I haven't yet tried, and wonder why it hasn't been done yet, is to put a small length of the gas-collecting tubing— very thin, narrow plastic or Teflon tubing—on the end of a pipet and bend it into a curve so you can go after your distillation product without resorting to glassworking. Again, the 9-in. pipet *can* just make it, but be very careful.

A Port in a Storm

Irritation, not necessity, is the real mother of invention, and irritation with getting the distillate out of the Hickman still led to a modification of the earlier design with a port on the side (Fig. 20.2). The addition of this side port makes removal of the distillate a lot easier.

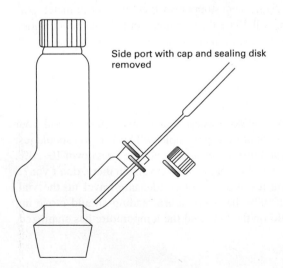

Side port with cap and sealing disk removed

FIGURE 20.2 Hickman still with port makes product removal easy.

THE ROTARY EVAPORATOR

If your product is dissolved in a solvent, you can distill off the solvent, boil off the solvent, or gently evaporate off the solvent using a **rotary evaporator**. Everybody calls it a **rotovap** and uses the word as a verb, too: "*Rotovap* the solution to dryness." Generally, they mean a rotary evaporator made by the Büchi Corporation, so you might be told to "put it on the Büchi."

The key to the rotovap is that it spins your flask, held at an angle, under a vacuum, and the solvent continually coats the upper part of the flask with a thin film. This thin film of solvent, covering such a large area of the flask, then evaporates pretty quickly.

From the top, the Büchi rotovap (Fig. 21.1) consists of the following:

1. *Solvent-sucking inlet.* If you have a lot of solution, you can suck more of it way down in the rotating flask without missing a beat. With a **splash trap** on the rotovap, this feature is useless, however. Generally, you leave this unconnected and use it to let air into the rotovap by turning the stopcock at the top.

2. *Stopcock at the top.* Note how the handle points to the flow of the liquid or gas. Handle away from the inlet tube, stopcock closed; handle on the same side as the inlet tube, stopcock open.

3. *Condenser.* This is a pretty fancy condenser in which the cooling water flows through this long, coiled glass tube inside the condenser. Cool! This alone can account for the virtual disappearance of other styles of rotovap that are just as effective and considerably cheaper. They are just not as cool. There's a glass inlet

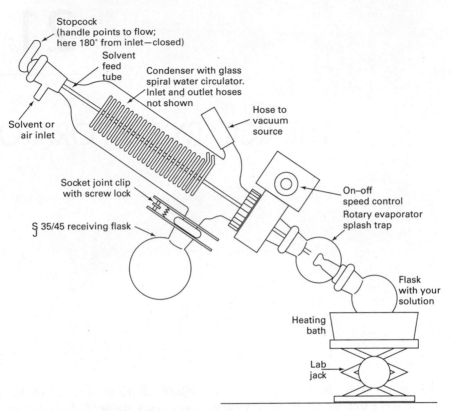

FIGURE 21.1   The Büchi rotary evaporator.

and outlet tube (not shown in Fig. 21.1) "around the back." They're close
enough to the vacuum source connection that you might at first confuse the
three, but a close inspection will reveal which is which.

4. **Vacuum connection.** This is where you, well, connect the vacuum source.

5. **Socket joint clip with screw lock.** This looks and acts a lot like a spring-type
clothespin. Just like that kind of clothespin, you press the ends together to open
the jaws of the clip and fit the jaws around the socket joint. Then, you turn a little
knurled wheel to lock the jaws around the joint. To remove the clip, you may
first have to unlock the jaws by unscrewing the little knurled wheel. If you don't
know about this, when you come to remove the receiving flask by pressing the
ends, the ends won't move. The first response here is not that the clip is broken,
but that you have to unscrew the little knurled wheel so the jaws can open.

6. **$35/25 receiving flask.** Unlike ⊤ joints that can only rotate about their axis,
ball-and-socket joints—or just socket joints—can rotate in almost any direction. An $35/25 joint has a diameter of 35 mm, with a bore of 25 mm going
through it.

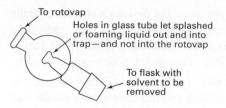

To rotovap

Holes in glass tube let splashed
or foaming liquid out and into
trap—and not into the rotovap

To flask with
solvent to be
removed

FIGURE 21.2   A rotovap splash guard.

7. *On–off speed control.* What more needs to be said? Don't immediately crank this control all the way up. You can get to speeds that seem scary.

8. *Rotovap splash trap* (Fig. 21.2). If your solution bumps or foams, you can contaminate the condenser with whatever was dissolved in the solvent. Hence the trap.

9. *Your flask.* Besides not having it more than half-full, there's nothing to say here but this: Yes, you can use clips to hold the flask, and even the splash trap, on the rotovap, but I've always let the vacuum in the setup hold the flask and trap on. Just lucky, I guess.

10. *Heating bath.* Just that. Unless the bath is thermostatted, it's probably better to just put warm water in the bath and let it cool as the evaporation continues. As the liquid evaporates from the walls of the flask, as they say in the chemistry books, "the fraction of molecules with the highest kinetic energy" leave, and the solution you're rotovapping cools, which means you have to add more heat to keep things evaporating at the same rate. If you care to, after you've started your rotovap but before you raise the heating bath, feel the rotating flask. It'll get cold.

11. *Lab jack.* Raises and lowers the heating bath.

Rotovapping is like vacuum distillation, and in starting a vacuum distillation it's always first the stirring, then the vacuum, then the heat. Here stirring is replaced by spinning the flask. To start rotovapping a solution:

1. Make sure the sample-sucking inlet stopcock is closed off. It usually is; just make sure. Empty the receiving flask. Suppose the last person rotovapped off diethyl ether (bp 34.6°C) and didn't clean it out. Then you come by and want to rotovap off ethanol (bp 78.5°C). If ether is still in the receiving flask, as you pull a vacuum in the rotovap, the ether will start to reflux, filling the rotovap with ether vapor. Your ethanol molecules haven't a chance. They'll just sit there. So, clean out the receiving flask. Reattach the socket joint flask to the ball joint on the condenser, and use the special clip to hold the flask to the condenser outlet. If you're doing solvent recovery so you can reuse it later, it's best to have a single rotovap dedicated to a single solvent. Otherwise, you can get cross-contamination of one solvent in another. Cleaning and drying the receiving flask may not be enough. There's a lot of surface area in that condenser.

2. Start the cooling water flowing through the condenser. Gentle flow, please.

3. Make sure your round-bottom jointware flask fits the rotovap or you have adapters to make it fit.

4. Slowly, without hooking anything up yet, angle the flask so it is parallel to the joint on the rotovap. Did some of your solution spill out? A clear indication that your flask is too full. Half-full is about the limit; a little less is better. Either switch to a bigger flask or you'll have to rotovap your solution in parts.

5. Without ever letting go of the flask (unless you've clipped it on), connect the flask to the joint on the rotovap. (Your flask might better have a splash trap connected to it, and *both* go onto the joint on the rotovap.) Still holding the flask, start the motor that spins the flask. Dial up a speed that you and your instructor are comfortable with.

6. Slowly establish a vacuum in the rotovap. When you are sure there's enough vacuum in the rotovap so the flask (and splash trap) won't drop off the beast, then, and only then, let go of the flask. Again, if you've clipped it on, you don't have to wait for the vacuum to build to hold the flask (and splash trap) on.

7. Jack up whatever heating bath is recommended, if any.

Now you let the flask spin and the solvent evaporate. It's a good idea to watch the setup every so often to see how far down the solvent is.

Again, rotovapping is like vacuum distillation, and in ending a vacuum distillation it's always first turn off the heat, then break the vacuum, then stop the stirring, and, again, stirring is replaced by spinning the flask. When you're ready to take the flask off the rotovap:

1. Lower or otherwise remove any heating bath.

2. Supporting the bottom of the flask (unless you've clipped it on), slowly let air into the rotovap; the sample-sucking inlet stopcock is the best place to do this.

3. Turn off the spinner motor.

4. Remove your flask.

5. Turn off whatever you used to establish a vacuum in the rotovap, either a water aspirator or a mechanical pump.

6. Turn off the cooling water.

7. Be a *mensch*, and empty the receiving flask for the next person. There might be a special container for these solvents. Check with your instructor.

8. The term *mensch* applies to persons of both genders. Look it up.

EXERCISES

1. Did you look up the Yiddish word *mensch*?

2. The last person to use the rotovap took off acetone. You need to take off ethanol. Should you check the receiving flask, and why should you? If a small amount of acetone is left in the receiving flask, is it a big deal or not? Why or why not?

REFLUX AND ADDITION

Just about 80% of the reactions in organic lab involve a step called **refluxing**. You use a reaction solvent to keep materials dissolved and at a constant temperature by boiling the solvent, condensing it, and returning it to the flask.

For example, say you have to heat a reaction to around 80°C for 17 hours. Well, you can stand there on your flat feet and watch the reaction all day. Me? I'm off to the **reflux**.

STANDARD REFLUX

Usually you'll be told what solvent to use, so selecting one should not be a problem. What happens more often is that you choose the reagents for your particular synthesis, put them into a solvent, and **reflux** the mixture. You boil the solvent and condense the solvent vapor *so that ALL of the solvent runs back into the reaction flask* (see "Class 3: Fractional Distillation" in Chapter 19). The *reflux temperature is near the boiling point of the solvent.* To execute a reflux,

1. Place the reagents in a round-bottom flask. The flask should be large enough to hold both the reagents and enough solvent to dissolve them, without being much more than half-full.

2. You should now choose a solvent that
   a. Dissolves the reactants at the boiling temperature.
   b. Does *not* react with the reagents.
   c. Boils at a temperature that is high enough to cause the desired reaction to go at a rapid pace.

3. Dissolve the reactants in the solvent. Sometimes *the solvent itself is a reactant.* Then don't worry.

4. To stop bumping, add a boiling stone to the flask. A magnetic stirring bar and stirrer will work, too.

5. Place a condenser, upright, on the flask, connect the condenser to the water faucet, and run water through the condenser (Fig. 22.1). Remember: in at the bottom and out at the top.

6. Put a suitable heat source under the flask, and adjust the heat so that the solvent condenses *no higher than halfway up the condenser.* You'll have to stick around and watch for a while, since this may take some time to get started. Remember that **reflux time** starts when the solvent in the flask starts condensing, and falling back into the flask, or **refluxing**. Warming the solvent, or seeing it boil in the flask without, well, refluxing, ain't a reflux. Once the reaction is stable, though, go do something else. You'll be ahead of the game for the rest of the lab.

7. Once this is going well, leave it alone until the reaction time is up. If it's an overnight reflux, wire the water hoses on so they don't blow off when you're not there.

8. When the reaction time is up, turn off the heat, let the setup cool, dismantle it, and collect and purify the product.

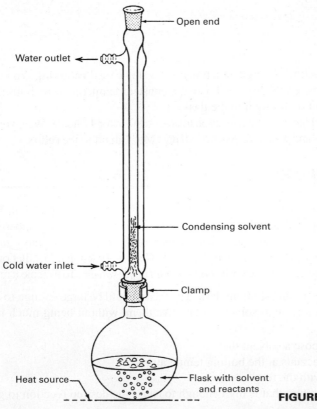

Open end

Water outlet

Condensing solvent

Cold water inlet

Clamp

Heat source

Flask with solvent and reactants

FIGURE 22.1 A reflux setup.

A DRY REFLUX

If you have to keep the atmospheric water vapor out of your reaction, you must use a **drying tube** and the **inlet adapter** in the reflux setup (Fig. 22.2). You can use these if you need to keep water vapor out of any system, not just the reflux setup.

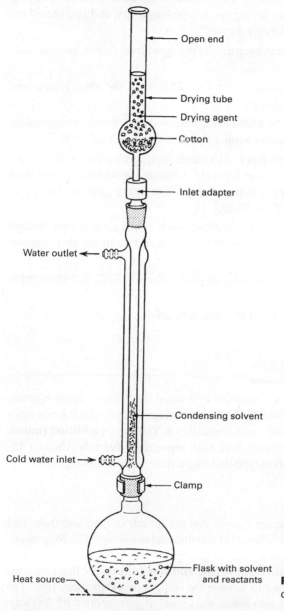

Open end

Drying tube

Drying agent

Cotton

Inlet adapter

Water outlet

Cold water inlet

Clamp

Heat source

Flask with solvent and reactants

FIGURE 22.2 Reflux setup à la drying tube.

1. If necessary, *clean and dry the drying tube.* You don't have to do a thorough cleaning unless you suspect that the anhydrous drying agent is *no longer anhydrous.* If the stuff is caked inside the tube, it is probably dead. You should clean and recharge the tube at the beginning of the semester. Be sure to use *anhydrous* calcium chloride or sulfate. It should last one semester. If you are fortunate, **indicating Drierite**, a specially prepared anhydrous calcium sulfate, might be mixed in with the white Drierite. If the color is *blue,* the drying agent is good; if it's *pink,* the drying agent is no longer dry, and you should get rid of it (see Chapter 10, "Drying Agents").

2. Put in a loose plug of *cotton* to keep the drying agent from falling into the reaction flask.

3. Assemble the apparatus as shown in Fig. 22.2, with the *drying tube and adapter on top of the condenser.*

4. At this point, reagents may be added to the flask and heated with the apparatus. Usually, the apparatus is heated while empty to drive water off its walls.

5. Heat the apparatus, usually empty, on a steam bath, giving the entire setup a quarter-turn every so often to heat it evenly. A burner can be used *if there is no danger of fire* and if heating is done carefully. The heavy ground-glass joints will crack if they are heated too much.

6. Let the apparatus cool to room temperature. As it cools, air is drawn through the drying tube before it hits the apparatus. The moisture in the air is trapped by the drying agent.

7. Quickly add the dry reagents or solvents to the reaction flask, and reassemble the system.

8. Carry out the reaction as usual, like a standard reflux.

ADDITION AND REFLUX

Every so often you have to add a compound to a setup while the reaction is going on, usually along with a **reflux**. Well, in adding new reagents, you *don't break open the system, let toxic fumes out, and make yourself sick.* You use an **addition funnel**. Now, we talked about addition funnels back with **separatory funnels** (Chapter 15) when we were considering the **stem**, and that might have been confusing.

Funnel Fun

Look at Figure 22.3*a*. It is a true sep funnel. You put liquids in here and shake and extract them. But could you use this funnel to add material to a setup? *NO*. No ground-glass joint on the end, and only glass joints fit glass joints. Right? Of course, right.

Figure 22.3*c* shows a **pressure-equalizing addition funnel**. See that side arm? Remember when you were warned to remove the stopper of a separatory funnel so that you wouldn't build up a vacuum inside the funnel as you emptied it? Anyway, the side arm equalizes the pressure on both sides of the liquid you're adding to the flask, so it'll flow freely without vacuum buildup and without any need for you to remove the stopper. This equipment is very nice, very expensive, very limited, and very rare. And if you try an **extraction** in one of these, all the liquid will bypass the stopcock and run out the tube onto the floor as you shake the funnel.

So a compromise was reached (Fig. 22.3*b*). Since you'll probably do more extractions than additions, with or without reflux, the pressure-equalizing tube is gone, but the ground-glass joint stayed on. Extractions? No problem. The nature of the stem is unimportant. But during additions, you'll have to take the responsibility to see that nasty vacuum buildup doesn't occur. You can remove the stopper every so often, or put a **drying tube** and an **inlet adapter** in place of the stopper. The inlet adapter keeps moisture out and prevents vacuum buildup inside the funnel.

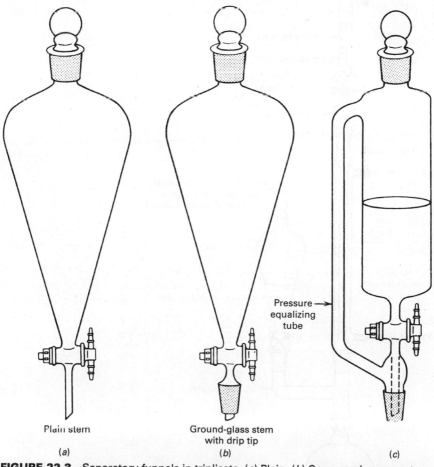

Pressure → equalizing tube

Plain stem

(*a*)

Ground-glass stem with drip tip

(*b*)

(*c*)

FIGURE 22.3 Separatory funnels in triplicate. (*a*) Plain. (*b*) Compromise separatory addition funnel. (*c*) Pressure-equalizing addition funnel.

How to Set Up

There are at least two ways to set up an addition and reflux, using either a **three-neck flask** or a **Claisen adapter**. I thought I'd show both of these setups with **drying tubes**. They keep the moisture in the air from getting into your reaction. If you don't need them, do without them.

Often the question comes up, "If I'm refluxing one chemical, how fast can I add the other reactant?" Try to follow your instructor's suggestions. Anyway, usually the reaction times are fixed. So I'll tell you what NOT, repeat NOT, to do.

If you reflux something, there should be a little ring of condensate, sort of a cloudy, wavy area in the barrel of the reflux condenser (Figs. 22.4 and 22.5). Assuming an exothermic reaction—the usual case—adding material from the funnel has the effect of heating up the flask. The ring of condensate begins to move *up*.

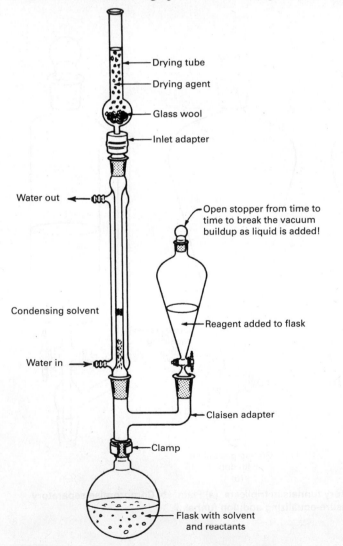

Drying tube
Drying agent
Glass wool
Inlet adapter
Water out
Open stopper from time to time to break the vacuum buildup as liquid is added!
Condensing solvent
Reagent added to flask
Water in
Claisen adapter
Clamp
Flask with solvent and reactants

FIGURE 22.4   Reflux and addition by Claisen tube.

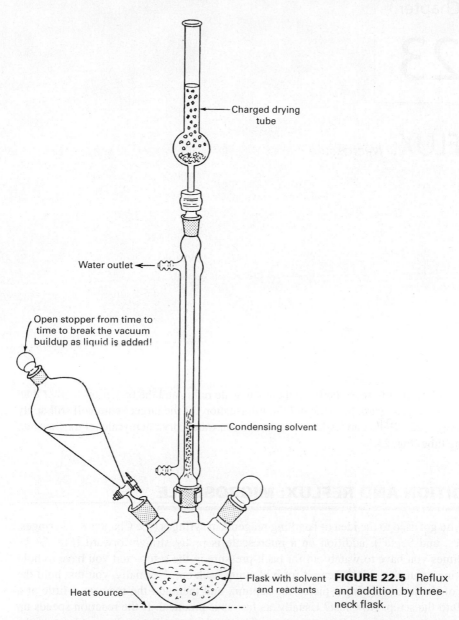

Charged drying
tube

Water outlet

Open stopper from time to
time to break the vacuum
buildup as liquid is added!

Condensing solvent

Heat source

Flask with solvent
and reactants

FIGURE 22.5 Reflux
and addition by three-
neck flask.

Well, don't *ever* let this get more than three-quarters up the condenser barrel. If the reaction is that fast, a very little extra reagent or heating will push that ring out of the condenser and possibly into the room air. No, no, no, no.

EXERCISE

If one of the reagents in a reaction you are using a reflux to carry out is not entirely soluble in the boiling solvent, can you still reflux? *Should* you still reflux? What considerations would come into play here?

REFLUX: Microscale

There's not much difference between the microscale reflux and the bigger ones other than the size of the glassware. So, much of the information for the larger setup will still apply (see Chapter 22, "Reflux and Addition"). And if you want a dry microscale reflux, then add a drying tube (Fig. 23.1).

ADDITION AND REFLUX: MICROSCALE

Once you get used to the idea of handling reagents in syringes (see Chapter 8, "Syringes, Needles, and Septa"), addition on a microscale is pretty straightforward (Fig. 23.2). Sometimes you have to watch out for back-pressure in the setup, and you have to hold the syringe plunger a bit to keep it from blowing back out. Normally, you just hold the barrel of the loaded syringe, pierce the septum, and then add the reagent, a little at a time, into the setup. How fast? Usually, as fast as is practical. If the reaction speeds up too much, the ring of condensate in the reflux condenser will travel up and out of the condenser and into the room air. No, no, no, no. This sounds identical to the warning I give for the large-scale setup. *Hint:* It is.

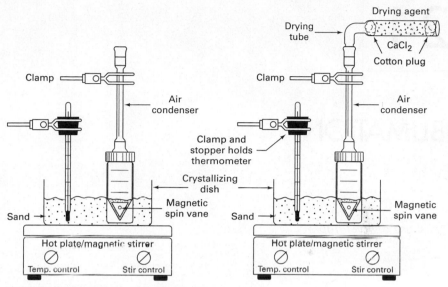

FIGURE 23.1 Microscale reflux, both wet and dry.

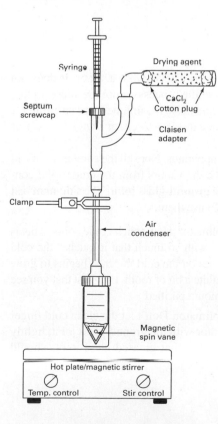

FIGURE 23.2 A dry microscale reflux and addition.

SUBLIMATION

- *You must use a water trap.*

Sublimation occurs when you heat a solid and it turns directly into a vapor. It does not pass GO, nor does it turn into a liquid. If you reverse the process—cool the vapor so that it turns back into a solid—you've *condensed the vapor*. Use the unique word **sublime** for the direct conversion of *solid to vapor*. **Condense** can refer to either *vapor-to-solid* or *vapor-to-liquid* conversions.

Figure 24.1 shows three forms of sublimation apparatus. Note all the similarities. Cold water goes in and down into a **cold finger** on which the vapors from the crude solid condense. The differences are that one is larger and has a **ground-glass joint**. The **side-arm test tube** with **cold-finger condenser** is much smaller. To use them,

1. Put the crude solid into the bottom of the sublimator. How much crude solid? This is rather tricky. You certainly don't want to start with so much that it touches the cold finger. And since, as the purified solid condenses on the cold finger, it begins to grow down to touch the crude solid, there has to be quite a bit of room. I suggest that you see your instructor, who may want only a small amount purified.

2. Put the cold finger into the bottom of the sublimator. Don't let the clean cold finger touch the crude solid. If you have the sublimator with the ground-glass joint, lightly (and I mean lightly) grease the joint. Remember that greased glass joints should NOT be clear all the way down the joint.

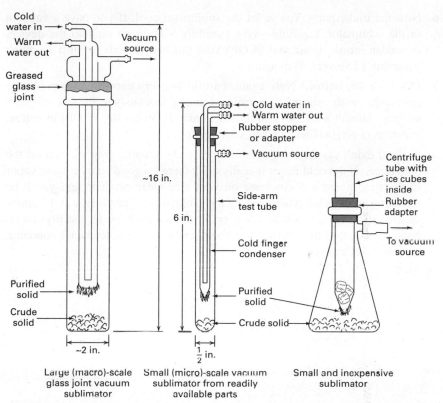

FIGURE 24.1 King-size, miniature, and inexpensive microscale sublimation apparatus.

3. Attach the hoses. *Cold water goes in the center tube*, pushing the warmer water out the side tube. Start the cooling water. Be careful! The inexpensive microscale sublimator needs no running cooling water—only ice and water in the tube.

4. If you're going to pull a vacuum in the sublimator, do it now. If the vacuum source is a **water aspirator**, *put a water trap between the aspirator and the sublimator*. Otherwise, you may get depressed if, during a sudden pressure drop, water backs up and fills your sublimator. Also, start the vacuum *slowly*. If not, air, entrained in your solid, comes rushing out and blows the crude product all over the sublimator, like popcorn.

5. When everything has settled down, slowly begin to heat the bottom of the sublimator, if necessary. You might see vapors coming off the solid. Eventually, you'll see crystals of *purified solid form on the cold finger*. Since you'll work with different substances, different methods of heating will have to be used. Ask your instructor.

6. Now the tricky parts. You've let the sublimator cool. If you have a vacuum in the sublimator, carefully—very carefully—introduce air into the device. A sudden inrush of air, and PLOP! Your purified crystals are just so much yesterday's leftovers. Start again.

7. **(And see #8, below.)** Now again, carefully—very carefully—remove the cold finger, with your pristine product clinging tenuously to the smooth glass surface, without a lot of bonking and shaking. Otherwise, PLOP! et cetera, et cetera, et cetera. Clean up and start again.

8. Notice I didn't say anything about immediately scraping your product off the cold finger. If the cold finger is really cold, there's a good chance water vapor in the atmosphere will condense on your nice clean product, and you'll be removing a wet solid. After opening the sublimator—*carefully*—to the atmosphere, turn off the cooling water, or let the ice melt, or let the dry-ice (!) bubble away until the product is at room temperature. Water won't condense on your solid anymore.

MICROSCALE BOILING POINT

The boiling point of a liquid is the temperature at which the vapor pressure of the liquid equals the atmospheric pressure. More practically, you see the liquid boil. The absolute best way to get the boiling point of a liquid is to distill it, and, unfortunately, as the bean counters exert more and more influence over the organic laboratory, there's less and less of a chance you'll ever get to determine a boiling point this way. If you're lucky enough to have this luxury, then go back and read the appropriate sections (see Chapter 19, "Distillation," and Chapter 20, "Microscale Distillation"). The rest of you poor devils will be using a technique developed years ago to determine the boiling point of a liquid on a scale of 2 to 5 traditional eyedropper drops.

I didn't have to distinguish drop size before the introduction of Pasteur pipets into the laboratory, but I do now. The drop size of a Pasteur pipet is very different from that of a drop from a traditional eyedropper. I took the average mass of 20 drops of room-temperature cyclohexanone and found that an eyedropper delivered about 0.0201 g and a Pasteur pipet delivered about 0.0137 g. In the following discussions, I'm using eyedropper drops, OK?

MICROSCALE BOILING POINT

Back in the old days, you would take a small (75 × 150-mm) test tube, strap it to a thermometer with a rubber band or ring, and put about 4–5 drops of your unknown liquid into the test tube. Then you'd get an ordinary melting-point capillary tube sealed

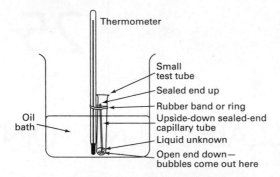

Thermometer

Small test tube
Sealed end up
Rubber band or ring
Upside-down sealed-end capillary tube
Liquid unknown
Open end down—bubbles come out here

Oil bath

FIGURE 25.1 Micro boiling-point apparatus.

on one end, drop the melting-point tube into the liquid in the test tube *open end down,* and clamp the whole shebang, so that the thermometer bulb, the test tube with your sample, and the open-end-down melting-point capillary in it were all in an oil bath (Fig. 25.1).

At this point, you'd be told to heat the bath and watch the open end of the capillary tube. As you heated your liquid unknown, first the air in the capillary would expand, and a very slow stream of bubbles would start to come out of the tube as the air left it. At some point—and I was never really sure about this point myself—the stream of bubbles was to get faster as the liquid started to boil. You'd let this go for a while and then remove the heat. Your boiling liquid would cool, and as soon as the liquid entered the capillary tube—BINGO!—you'd read the thermometer and get the first value of the boiling range. When the liquid finally filled the tube, you'd read the temperature again and get the second value of the boiling range. There were, of course, a few things to watch out for:

1. *When do the air bubbles become sample vapor bubbles?* Good question, and the one that still bothers me the most. Ideally, you won't rush heating the sample, and there will be a noticeable difference between the rate of formation of air expulsion bubbles and actual boiling bubbles. If your sample boils at a relatively high temperature (probably >100°C), you have more time to heat it and convince yourself that the bubbles you see are those of your boiling liquid and not just air trying to leave the tube. Try this with, say, acetone (bp 56°C), and you could boil away all of the liquid faster than you can decide if the liquid is boiling away.

   If you're sure your liquid doesn't decompose, react, or change in some way when boiling, you can stop the heating after what you believe to be the rapid-boiling bubbles appear. When that sample cools and the liquid rushes back into the tube and fills it completely, you know those bubbles were not air, but boiling sample. The advantage you have now is that when you reheat the sample, any bubbles you see can't be air bubbles; they have to be boiling sample bubbles.

Now when you reheat the sample and the stream of bubbles starts again, you can be confident that the bubbles are not air, but your boiling sample. Let the sample cool, and the temperatures you record are the boiling range, called boiling point, of your liquid sample. Again, the disadvantage is that the sample may decompose or in some way change when boiling, and your data are meaningless.

2. *I only got one reading on my thermometer.* A melting point can appear to be one temperature, rather than a range, however small, if you heat the sample too rapidly. So also for this method of boiling-point determination, you can appear to get one temperature, rather than a range, however small, if you cool the sample too rapidly.

ULTRAMICROSCALE BOILING POINT

The only modification, if you can call it that, in the ultramicroscale technique is to perform it on an even smaller scale. The test tube into which you put your sample becomes a melting-point capillary, and the closed-end-up, open-end-down melting-point capillary is a smaller version you have to make yourself! Out of another melting-point capillary, no less.

Heat the midsection of an open-end melting-point capillary tube, and carefully draw the glass to form an even smaller capillary section. Break this section in two—just as if you were making a capillary spotter (see Chapter 27, "Thin-Layer Chromatography"). At this point, you seal the narrow end on each "capillary spotter" and then break this narrow tube off the original, unchanged larger section of the melting-point capillary tube. Violà—two itty-bitty, really narrow capillary tubes. Put these tubes, which are about 5 to 8 mm long, into the bigger melting-point capillary, open end down and closed end up. If this seems familiar, well, it should. It's a smaller version of the test-tube-and-melting-point-capillary setup described earlier. And you heat this by putting it in a melting-point apparatus! (See Chapter 12, "The Melting-Point Experiment.") As usual, in addition to the same problems of determining the boiling point that you've seen for the larger setup above, this one has its own set of problems:

1. *Loading the melting-point capillary with sample.* The only practical way to do this is with a syringe. You load a syringe with your liquid and squirt it into the capillary tube (see Chapter 8, "Syringes, Needles, and Septa"). The needles are usually too short to place the liquid directly at the bottom of the melting-point capillary, so you'll probably have to centrifuge the tube to force the liquid to the bottom.

2. *Making the itty-bitty, really narrow capillary not narrow enough.* If this is the case, then your liquid sample will wick up between the walls of the melting-point capillary and your handmade capillary, and you might not have a big

enough pool of liquid at the bottom of the melting-point capillary for those bubbles to form.

3. *Making the itty-bitty, really narrow capillary not heavy enough.* If your hand-made capillary is too light (usually from being too short), then it can float around in the melting-point capillary and give results that might be difficult to interpret. One way of making sure that your homemade capillary is heavy enough (besides making it overly long—don't do this! Anything about 5 to 8 mm long is just about right!) is to make sure you leave a bit of fused glass on the end to help weigh the homemade capillary down. Be careful, though. If you get too creative and leave a really unusual glob of glass on the end, your homemade capillary might get stuck at the top of the melting-point capillary.

Well, that's about the end of the ultramicroscale boiling-point technique, but for the fact that you still have the same problems as with the microscale boiling-point technique, already described.

CHROMATOGRAPHY: Some Generalities

Chromatography is the most useful means of separating compounds to purify and identify them. Indeed, separations of colored compounds on paper strips gave the technique its colorful name. Although there are many different types of chromatography, all the forms bear tremendously striking similarities. **Thin-layer**, **wet-column**, and **dry-column chromatography** are common techniques that you'll run across.

This chromatography works by differences in **polarity**. (That's not strictly true for all types of chromatography, but I don't have the inclination to do a 350-page dissertation on the subject, when all you might need to do is separate the differently colored inks in a black marker pen.)

ADSORBENTS

The first thing you need is an **adsorbent**, a porous material that can suck up liquids and solutions. Paper, silica gel, alumina (ultrafine aluminum oxide), cornstarch, and kitty litter (unused) are all fine adsorbents. Only the first *three* are used for chromatography, however. You may or may not need a **solid support** with these. Paper hangs together, is fairly stiff, and can stand up by itself. Silica gel, alumina, cornstarch, and kitty litter are more or less powders and need a solid support such as a glass plate to hold them.

Now you have an *adsorbent on some support*, or a *self-supporting adsorbent*, such as a strip of paper. You also have a mixture of stuff you want to separate. So you dissolve the mixture in an easily evaporated solvent, such as methylene chloride, and put some of it on the adsorbent. *Zap!* It is adsorbed! Stuck on and held to the adsorbent. But because you have a mixture of different things and they are different, they will be *held on the adsorbent in differing degrees*.

SEPARATION OR DEVELOPMENT

Well, so there's this mixture, sitting on this adsorbent, looking at you. Now you start to *run solvents through the adsorbent*. Study the following list of solvents. Chromatographers call these solvents **eluents** (or **eluants**).

THE ELUATROPIC SERIES

Not at all like the World Series, **the eluatropic series** is simply a list of solvents arranged according to increasing polarity.

Some solvents arranged
in order of increasing polarity

(Least polar)	Petroleum ether
	Cyclohexane
Increasing	Toluene
polarity	Chloroform
	Acetone
	Ethanol
(Most polar) ↓	Methanol

So you start running "petroleum ether" (remember, pet. ether is a mixture of hydrocarbons like gasoline—it's not a true ether at all). It's not very polar. So it is *not held strongly to the adsorbent*.

Well, this solvent is traveling through the adsorbent, minding its own business, when it encounters the mixture placed there earlier. It tries to kick the mixture out of the way. But most of *the mixture is more polar, held more strongly* on the adsorbent. Since the pet. ether cannot kick out the compounds that are more polar than itself very well, most of *the mixture is left right where you put it*.

No separation.

Desperate, you try methanol, one of the most polar solvents. It is *really held strongly to the adsorbent*. So it comes along and kicks the living daylights out of just about *all* of the molecules in the mixture. After all, the methyl alcohol *is more polar,*

200

so it can move right along and displace the other molecules. And it does. So, when you evaporate the methanol and look, *all the mixture has moved with the methanol,* so you get *one spot* that moved right with the **solvent front**.

No separation.

Taking a more reasonable stand, you try chloroform, because it has an intermediate polarity. The chloroform comes along, sees the mixture, and is able to push out, say, *all but one* of the components. As it travels, kicking the rest along, it gets tired (the eluent is not polar enough) and starts to leave some of the more polar components behind. After a while, *only one component is left moving with the chloroform,* and that may be dropped, too. So, at the end, *there are several spots left,* and each of them is in a *different place* from where it was at the start. Each spot is *at least one different component* of the entire mixture.

Separation. At last!

You could also use a solvent mixture if none of these pure solvents gives you the separation you want. Unfortunately, the effects don't just add up: A little bit of a polar solvent goes a long way. A 50–50 mixture of any two solvents is going to wind up with a polarity closer to the more polar solvent, *not* in the middle at all.

I picked these solvents for illustration. They are quite commonly used in this technique. I worry about the hazards of using chloroform, however, because it's been implicated in certain cancers. Many other common solvents are also suspected carcinogens. In the lab, either you will be told what solvent (eluent) to use or you will have to find out yourself, mostly by trial and error.

THIN-LAYER CHROMATOGRAPHY: TLC

Thin-layer chromatography (TLC) is used for identifying compounds and determining their purity. You can even follow the course of a reaction by using the technique. In short, you place a spot of the substance on the adsorbent surface of a TLC plate, let a solvent (or solvents) run up through the adsorbent, and visualize the plate to see what, if anything, happened to your compound (see Chapter 26, "Chromatography: Some Generalities").

WE DON'T MAKE OUR OWN
TLC PLATES ANY MORE, BUT...

We used to make TLC plates on 1 in. × 3 in. microscope slides, which were convenient, inexpensive, and fitted the most popular of beakers as development chambers. Now, we simply purchase them pre-made (see below for more) and have them cut to 25 × 75 mm for some reason.

Silica gel and alumina are popular column chromatography adsorbents, and if you're using TLC to see how your compound will behave on a column, then the adsorbent on the TLC plate should match the adsorbent you're using in the column. If you're not going on to column chromatography, then the choice of adsorbent isn't as important.

Since you're not making the TLC plates, you won't need to know that the container marked "Silica Gel G 254" has a G for gypsum binder to stick to the plates, is a compound that fluoresces at UV 254 nm for visualization of plates, and is not for column chromatography.

Pre-prepared TLC Plates

Pre-prepared TLC plates have an active coating, usually on a thin plastic sheet, also with or without the fluorescent indicator. Standard sizes are 5 × 20 cm and 20 × 20 cm, so they have to be cut using a pair of scissors—probably to 25 × 75 mm—before you can use them in the undergraduate lab. *Don't touch the active surface with your fingers— handle them only by the edge.* The layer on the plate is pretty thin, and bits can flake off at the edge if you're not careful. The thinness of the layer also means that you have to use very small amounts of your compounds in order not to overload the adsorbent.

THE PLATE SPOTTER

1. The **spotter** is the apparatus used to put the solutions that you want to analyze on the plate. You use it to make a spot of sample on the plate.

2. Put the center of a **melting-point capillary** into a small, blue Bunsen burner flame. Hold it there until the tube softens and starts to sag. Do not rotate the tube, ever.

3. *Quickly* remove the capillary from the flame, and pull both ends (Fig. 27.1). If you leave the capillary in the flame too long, you get an obscene-looking mess.

4. Break the capillary at the places shown in Figure 27.1 to get **two spotters** that look roughly alike. (If you've used capillaries with *both ends open already,* then you don't have a closed end to break off.)

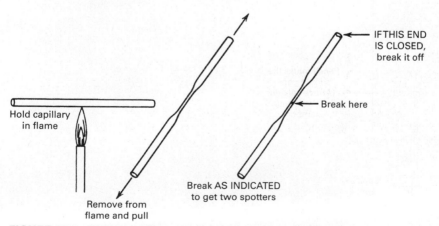

FIGURE 27.1 Making capillary spotters from melting-point tubes.

5. Make up 20 of these, or more. You'll need them.

6. Because TLC is so sensitive, spotters tend to "remember" old samples if you reuse them. *Don't put different samples in the same spotter.*

SPOTTING THE PLATES

Don't spot the shiny side; that's the plastic not the absorbent!

1. You might want to lightly mark the place where you're going to spot a plate with a pencil (never ink! *Never!*), especially if your sample is dilute and you have to spot more than once in the sample place (Fig. 27.2).

2. Dissolve a small portion (1–3 mg) of the substance you want to chromatograph in *any* solvent that dissolves it *and* evaporates rapidly. Dichloromethane or diethyl ether often works best.

3. Put the thin end of the capillary spotter into the solution. The solution should rise up into the capillary.

4. Touch the capillary to the plate *briefly!* The compound will run out and form a small spot. Try to keep the spot as small as possible—not larger than $\frac{1}{4}$ in. in diameter. Blow gently on the spot to evaporate the solvent. Touch the capillary to the *same place*. Let this new spot grow to be *almost the same size as the*

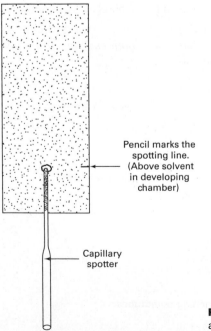

Pencil marks the spotting line. (Above solvent in developing chamber)

Capillary spotter

FIGURE 27.2 Putting a spot of compound on a TLC plate.

one already there. Remove the capillary and evaporate the solvent by gently blowing on the spot. This will build up a concentration of the compound.

5. You've got a bit of a choice here: Either lightly draw *in pencil* a line all the way across the plate about 1 cm from the end **now**, and stop the development when the solvent gets up to it, or develop the plate first, and as soon as you think it's done, pull it out of the development chamber and draw a line *in pencil* where that solvent front has gotten up to. Since solvents have no concept of time, I usually choose the latter, and, when I feel the plate has developed enough, pull the plate and draw the line. Take a bit of care; you'll be taking measurements from this line.

DEVELOPING A PLATE

1. Take a 150-mL beaker, line the sides of it with filter paper, and cover with a watch glass (Fig. 27.3). Yes, you can use other containers. I've used beakers because all chemistry stockrooms have beakers, you know what beakers look like, and it's a handy single word for the more cumbersome "thin-layer development chamber used at your institution." So if you do use jelly jars, drinking glasses, wide-mouth screwcap bottles, or whatever, just substitute that description for beaker in this discussion, OK?

2. Choose a solvent to **develop** the plate. You let this solvent (eluent) pass through the adsorbent by capillary action. Nonpolar eluents (solvents) force nonpolar compounds to the top of the plate, whereas polar eluents force *both polar and nonpolar materials up the plate*. There is only one way to choose eluents: educated guesswork. Use the chart of eluents presented in Chapter 26.

3. Pour some of the eluent (solvent) into the beaker, and tilt the beaker so that the solvent wets the filter paper. Put no more than $\frac{1}{4}$ in. of eluent in the bottom of the beaker! This saturates the air in the beaker with eluent (solvent) and stops the evaporation of eluent from the plate.

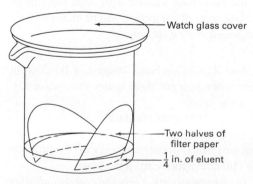

Watch glass cover

Two halves of filter paper

$\frac{1}{4}$ in. of eluent

FIGURE 27.3 The secret identity of a 150-mL beaker as a TLC slide development chamber is exposed. You can substitute a jar with a cap.

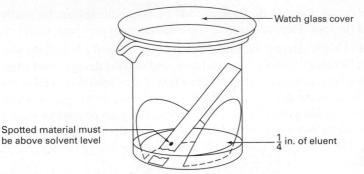

Watch glass cover

Spotted material must be above solvent level

$\frac{1}{4}$ in. of eluent

FIGURE 27.4 Prepared, spotted TLC plate in a prepared developing chamber. You can substitute a jar with a cap.

4. Place the slide in the developing chamber as shown (Fig. 27.4). *Don't let the solvent in the beaker touch the spot on the plate,* or the spot will dissolve away into the solvent! If this happens, you'll need a new plate, and you'll have to clean the developing chamber as well.

5. Cover the beaker with a watch glass. The solvent (eluent) will travel up the plate. The filter paper keeps the air in the beaker saturated with solvent, so that it doesn't evaporate from the plate. When the solvent reaches the line, *immediately* remove the plate. Drain the solvent from it, and immediately mark—*in pencil*—the solvent front. Then you can blow gently on the plate until *all* of the solvent is gone. If not, there will be some trouble visualizing the spots.

Don't breathe in the fumes of the eluents! Make sure you have adequate ventilation. Work in a hood if possible.

VISUALIZATION

Unless the compound is colored, the plate will be blank, and you won't be able to see anything, so *you must visualize the plate.*

1. *Destructive visualization.* Spray the plate with sulfuric acid, then bake in an oven at 110°C for 15–20 minutes. Any spots of compound will be charred blots, utterly destroyed. *All* spots of compound will be shown.

2. *Nondestructive visualization.*
   a. *Long-wave UV (Hazard!).* Most TLC adsorbents contain a fluorescent powder that glows bright green when you put them under long-wave UV light. There are two ways to see the spots:
      (1) The background glows green, and the spots are dark.
      (2) The background glows green, and the spots glow some other color. The presence of *excess eluent may cause whole sections of the plate to remain dark.* Let all of the eluent evaporate from the plate.
   b. *Short-wave UV (Hazard!).* The plates stay dark. *Only the compounds may glow.* This is usually at 180 nm.

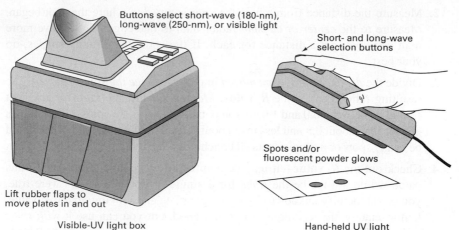

Buttons select short-wave (180-nm),
long-wave (250-nm), or visible light

Short- and long-wave
selection buttons

Spots and/or
fluorescent powder glows

Lift rubber flaps to
move plates in and out

Visible-UV light box

Hand-held UV light

FIGURE 27.5   Visible-UV light box and hand-held UV light.

Both of the UV tests can be done in a matter of seconds in a **UV light box** (Fig. 27.5). Since most compounds are unchanged by exposure to UV light, the test is considered **nondestructive**. Not everything will show up, but the procedure is good enough for most compounds. When using the light box, *always turn it off when you leave it*. If you don't, not only does the UV filter burn out, but also your instructor becomes displeased.

Since neither the UV nor the iodine test is permanent, it helps to have a record of what you've seen. You must *draw an accurate picture of the plate in your notebook*. Using a sharp-pointed object (pen point, capillary tube, etc.), you can trace the outline of the spots on the plate while they are under the UV light (**CAUTION!** Wear gloves!) or before the iodine fades from the plate.

3. *Semidestructive visualization.* Set up a developing tank (150-mL beaker) but leave out the filter paper and any solvent. Just a beaker with a cover. Add a few crystals of **iodine**. Iodine vapors will be absorbed onto most spots of compound, coloring them. Removing the plate from the chamber causes the iodine to evaporate from the plate, and *the spots will slowly disappear. Not all spots may be visible.* So if there's nothing there, that doesn't mean nothing's there. The iodine might have reacted with some spots, changing their composition. Hence the name **semidestructive visualization**.

INTERPRETATION

After visualization, there will be a spot or spots on the plate. Here is what you do when you look at them.

1. Measure the distance *from that solvent line* drawn across the plate *to where the spot started*.

2. Measure the distance from where the *spot stopped* to where the spot began. Measure to the *center of the spot* rather than to one edge. If you have more than one spot, get a distance for each. If the spots have funny shapes, do your best.

3. Divide the *distance the solvent moved* into the *distance the spot(s) moved*. The resulting ratio is called the R_f **value.** *Mathematically,* the ratio for any spot should be between 0.0 and 1.0, or you goofed. *Practically,* spots with R_f values greater than about 0.8 and less than about 0.2 are hard to interpret. They could be single spots or multiple spots all bunched up and hiding behind one another.

4. Check out the R_f **value**—it may be helpful. In identical circumstances, this value would always be the *same* for a single compound. If this were true, you could identify unknowns by running a plate and looking up the R_f value. Unfortunately, the technique is not that good, but you can use it *with some judgment* and a *reference compound* to identify unknowns (see "Multiple Spotting," following).

Figures 27.6 to 27.8 provide some illustrations. Look at Figure 27.6: If you had a mixture of compounds, you could never tell. This R_f value gives no information. Run this compound again. Run a new plate. *Never redevelop an old plate!*

Use a more polar solvent!

No information is in Figure 27.7, either. You couldn't see a mixture if it were here. Run a new plate. *Never redevelop an old plate!*

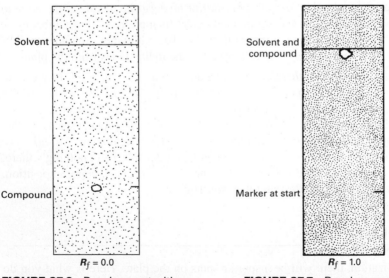

FIGURE 27.6 Development with a nonpolar solvent and no usable results.

FIGURE 27.7 Development with a very polar solvent and no usable results.

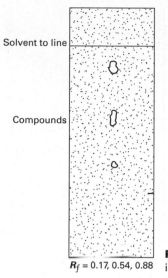

Solvent to line

Compounds

$R_f = 0.17, 0.54, 0.88$

FIGURE 27.8 Development with just the right solvent is a success.

Use a less polar solvent!

If the spot moves somewhere between the two limits (as shown in Fig. 27.8) and *remains a single spot*, the compound is *probably* pure. You should try a second solvent system, or a second adsorbent, before drawing this conclusion. If *more than one spot shows,* the compound is definitely impure, and it is a mixture. Whether the compound should be purified is a matter of judgment.

MULTIPLE SPOTTING

You can run more than one spot, either to save time or to make comparisons. You can even *identify unknowns*.

Let's say that there are two unknowns, A and B. Say one of them can be biphenyl (a colorless compound that smells like mothballs). You spot two plates, one with A and biphenyl side by side, and the other with B and biphenyl side by side. After you develop both plates, you have the results shown in Fig. 27.9.

Apparently, A is biphenyl.

Note that the R_f values are not perfect. This is an imperfect world, so don't panic over a slight difference.

Thus, we have a method that can quickly determine:

1. Whether a compound is a mixture.
2. The identity of a compound *if a standard is available.*

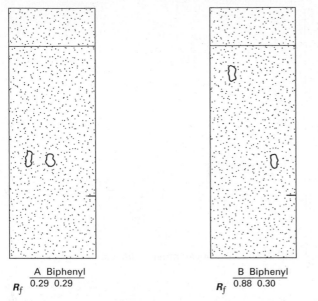

A Biphenyl
R_f 0.29 0.29

B Biphenyl
R_f 0.88 0.30

FIGURE 27.9   Side-by-side comparison of an unknown and a leading brand known.

COSPOTTING

You can nail down identification with **cospotting**, but it is a bit tricky. This time, spot one plate with your unknown, A, in two places. Now let the spots dry entirely. I mean entirely! Now we'll spot biphenyl *right on top of one of the spots of A*. (If you didn't let the earlier spot dry entirely, the biphenyl spot would bleed and you'd have a *very large*—and useless—spot of mixed compounds.) Do the same for B on another plate. Now run these plates.

The R_f values might be a bit different, but there's only *one spot* in the biphenyl over the A column. No separation. They are the same. If you see two (or more) spots come from cospotting, the substances, like B, are not the same (Fig. 27.10).

OTHER TLC PROBLEMS

What you've seen is what you get if everything works OK. You can:

1. Make the spots too concentrated. Here spots smear out, and you don't see any separation. Dilute the spotting solution, or don't put so much on (Fig. 27.11).
2. Put the spots too close together. Here you can get the spots bleeding into each other, and you might not be able to tell which spot came from which origin (Fig. 27.12).

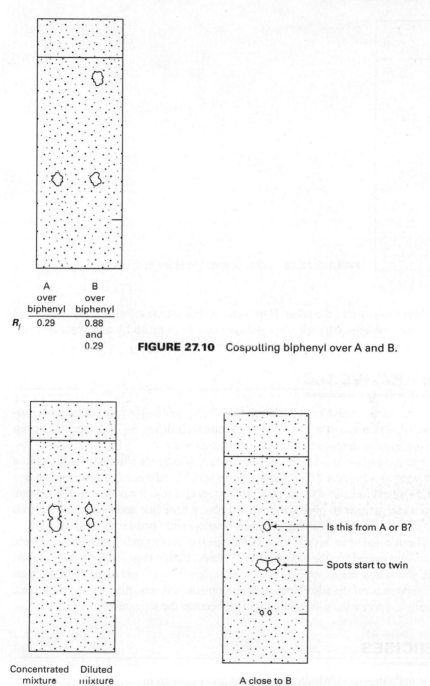

	A over biphenyl	B over biphenyl
R_f	0.29	0.88 and 0.29

FIGURE 27.10   Cospotting biphenyl over A and B.

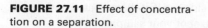

Concentrated mixture   Diluted mixture

FIGURE 27.11   Effect of concentration on a separation.

Is this from A or B?

Spots start to twin

A close to B

FIGURE 27.12   Effect of spots being too close together.

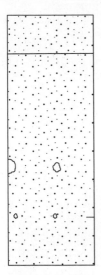

FIGURE 27.13   Effect of spot too close to the edge.

3. Spot too close to the edge. Here you get inaccurate R_f values, bleeding, and other problems. The spots are not surrounded as much by adsorbent and solvent, so unequal forces are at work here (Fig. 27.13).

PREPARATIVE TLC

When you use an **analytical technique** (like TLC) and you expect to **isolate compounds**, it's often called a **preparative (prep) technique**. So TLC becomes **prep TLC**. You use the same methods, only on a larger scale.

You usually use a 20 × 20 cm glass, heavy plastic, or aluminum plate, with a thicker layer of adsorbent (0.5–2.0 mm). Years ago, I used a small paintbrush to put a line of dissolved sample (a streak rather than a spot) across the plate near the bottom. Now you can get special plate streakers that give a finer line and less spreading. You put the plate in a large developing chamber, develop, and visualize the plate as usual.

The thin line separates and spreads into bands of compounds, much like a tiny spot separates and spreads on the analytical TLC plates. Rather than just look at the bands, though, you *scrape the adsorbent holding the different bands into different flasks,* blast your compounds off the adsorbents with appropriate solvents, filter off the adsorbent, and finally evaporate the solvents and actually recover the separate compounds.

EXERCISES

1. What's the matter? Ballpoint pen not good enough to mark a TLC plate?
2. A TLC plate has a shiny side and a dull side (as if it were coated with some kind of powder). Which should you use to spot your samples on?

3. What are the ways to visualize a TLC plate? Compare, contrast, and all that.

4. Your TLC plate has just finished its run. Why could the entire plate be dark at both wavelengths in the UV light box?

5. Define R_f value.

6. Compounds on TLC plates separate by molecular weight, don't they? Explain.

7. Two spots on a TLC plate have slightly different R_f values. Could they still be the same compound? Explain. How would you spot a new plate to remove all doubt?

8. Describe the effects of the following conditions: Samples are spotted too low on the TLC plate; eluent is too polar; eluent is not polar enough; sample is too concentrated; the top is left off the TLC chamber; the plate is left in the chamber after the solvent has reached the top of the plate.

9. You know there are organic acids on your TLC plates and they are tailing badly. Causes and cures? Is there really any difference in the characteristics of acid-washed alumina and base-washed silica gel?

Chapter 28

WET-COLUMN CHROMATOGRAPHY

Wet-column chromatography, as you may have guessed, is chromatography carried out on a **column of adsorbent** rather than a layer. It is cheap, easy, and carried out at room temperature, and you can also separate large amounts, *gram quantities,* of mixtures.

In column chromatography, the adsorbent is either alumina or silica gel. Alumina is basic and silica gel is acidic. If you try out an eluent (solvent) on silica gel plates, you should use a silica gel adsorbent. And if you have good results on alumina TLC (thin-layer chromatography), use an alumina column.

Now you have a *glass tube as the support* that holds the adsorbent in place. You dissolve your mixture and put it on the adsorbent at the top of the column. Then you wash the mixture down the column using at least one eluent (solvent), perhaps more. The compounds carried along by the solvent are washed *entirely out of the column,* into separate flasks. Then you isolate the separate fractions.

PREPARING THE COLUMN

1. The adsorbent is supported by a glass tube that has either a stopcock or a piece of tubing and a screw clamp to control the flow of eluent (Figs. 28.1 and 28.2). You can use an ordinary buret. What you will use will depend on your own lab program. Right above this control, you put a wad of cotton or glass wool to keep everything from falling out. *Do not use too much cotton or glass wool,* and

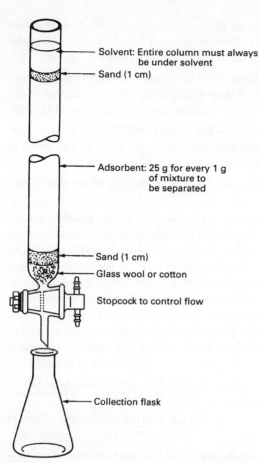

Solvent: Entire column must always
be under solvent

Sand (1 cm)

Adsorbent: 25 g for every 1 g
of mixture to
be separated

Sand (1 cm)

Glass wool or cotton

Stopcock to control flow

Collection flask

FIGURE 28.1 Wet-column chromatography setup.

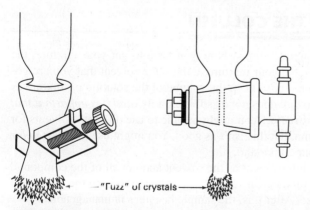

"Fuzz" of crystals

FIGURE 28.2 A growth of crystals occurs as the eluent evaporates.

do not pack it too tightly. If you ram the wool into the tube, the flow of eluent will be very slow, and you'll be in the lab until next Christmas waiting for the eluent. If you pack it *too loosely,* all the stuff in the column *will fall out.*

2. At this point, fill the column halfway with the least polar eluent you will use. If this is not given, you can guess it from a quick check of the separation of the mixture on a TLC plate. This would be the advantage of an alumina TLC plate.

3. Slowly put sand through a funnel into the column until there is a 1-cm layer of sand over the cotton. Adsorbent alumina or silica gel is so fine that either is likely to go through cotton but not through a layer of fine sand.

4. During this entire procedure, *keep the level of the solvent above that of any solid material in the column!*

5. Now slowly add the adsorbent. There are a few ways to do this: ***powder packing*** and ***slurry packing***. Alumina and silica gel are adsorbents, and when they suck up the solvent, they liberate heat. With powder packing, if you add the adsorbent too quickly, the solvent may boil and ruin the column. So don't add it too fast; add the adsorbent slowly! With slurry packing, you add the solvent to the adsorbent in a beaker, let the adsorbent do its exothermic thing, and pour the slurry into the column. In any case, use about 25 g of adsorbent for every 1 g of mixture you want to separate. While adding the adsorbent, tap or gently swirl the column to dislodge any adsorbent or sand on the sides. You know that a plastic wash bottle with eluent in it can wash the stuff down the sides of the column very easily.

6. When the alumina or silica gel settles, you normally have to add sand (about 1 cm) to the top to keep the adsorbent from moving around.

7. Open the stopcock or clamp, and let solvent out until the level of the solvent is just above the upper level of sand.

8. *Check the column! If there are air bubbles or cracks in the column, dismantle the whole business and start over!*

COMPOUNDS ON THE COLUMN

If you've gotten this far, congratulations! Now you have to get your mixture, the analyte, on the column. Dissolve your mixture in the same solvent that you are going to put through the column. Try to keep the volume of the solution or mixture as *small as possible.* If your mixture does not dissolve entirely (*and it is important that it do so*), check with your instructor! You might be able to use different solvents for the analyte and for the column, but this isn't as good. You might use the *least polar* solvent that will dissolve your compound.

If you must use the column eluent as the solvent and not all of the compound dissolves, you can filter the mixture through filter paper. Try to keep the volume of solution down to 10 mL or so. After this, the sample becomes unmanageable.

1. Use a pipet and rubber bulb to slowly and carefully add the sample to the top of the column (Fig. 28.3). *Do not disturb the sand!*

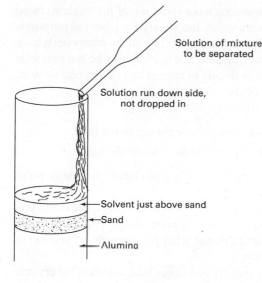

Solution of mixture
to be separated

Solution run down side,
not dropped in

Solvent just above sand

Sand

Alumina

FIGURE 28.3 Putting compounds on the column by pipet.

2. Open the stopcock or clamp, and let solvent flow out until the level of the solution of compound is slightly above the sand. *At no time let the solvent level get below the top layer of sand!* The compound is now "on the column."

3. Now add eluent (solvent) to the column above the sand. *Do not disturb the sand!* Open the stopcock or clamp. Slowly let eluent run through the column until the first compound comes out. Collect the different products in Erlenmeyer flasks. You may need lots and lots of Erlenmeyer flasks. *At no time let the level of the solvent get below the top of the sand!* If necessary, stop the flow, add more eluent, and start the flow again.

VISUALIZATION AND COLLECTION

If the compounds are colored, you can watch them travel down the column and separate. If one or all are colorless, you have problems. So,

1. *Occasionally* let one or two drops of eluent fall on a clean glass microscope slide. Evaporate the solvent and see if there is any sign of your crystalline compound! This is an excellent spot test, but don't be confused by nasty plasticizers from the tubing (Fig. 28.2) that try to put one over on you by pretending to be your product.

2. Put the narrow end of a "TLC spotter" to a drop coming off at the column. The drop will rise up into the tube. Using this loaded spotter, *spot, develop, and visualize a TLC plate with it.* Not only is this more sensitive, but you can also see whether the stuff coming out of the column is pure (see Chapter 27, "Thin-Layer Chromatography"). You'll probably have to collect more than one drop on a TLC plate. If it is very dilute, the plate will show nothing, even if there actually is compound there. It is best to sample four or five consecutive drops.

Once the first compound or compounds have come out of the column, those that are left may move down the column much too slowly for practical purposes. Normally you start with a nonpolar solvent. But by the time all the compounds have come off, it may be time to pick up your degree. The solvent may be too nonpolar to kick out later fractions. So you have to decide to change to a more polar solvent. This will kick the compound right out of the column.

To change solvent in the middle of a run:

1. Let the old solvent level run down to just above the top of the sand.

2. *Slowly add new,* **more polar solvent**, *and do not disturb the sand.*

You, and you alone, have to decide if and when to change to a more polar solvent. (Happily, sometimes you'll be told.)

If you have only two components, start with a nonpolar solvent, and when you are sure the *first component is completely off the column, change to a really polar one.* With only two components, it doesn't matter what polarity solvent you use to get the second compound off the column.

Sometimes the solvent evaporates quickly and leaves behind a "fuzz" of crystals around the tip (Fig. 28.2). Just use some fresh solvent to wash them down into the collection flask.

Now all the components are off the column and in different flasks. Evaporate the solvents (*no flames!*), and lo! The crystalline material is left.

Dismantle the column. Clean up. Go home.

WET-COLUMN CHROMATOGRAPHY: MICROSCALE

Occasionally, you'll be asked to clean and/or dry your product using **column chromatography** by pipet. It's a lot like large-scale, wet-column chromatography.

One problem with this pipet column is that there's no stopcock. Once it starts running, it goes until it runs out. And you can't let the column dry out, either. So it looks like you're committed to running the column once you prepare it.

In general (Fig. 28.4),

1. Shorten the tip of a Pasteur pipet (see "Pipet Cutting" in Chapter 7, "Pipet Tips") and tamp cotton into the tip as usual.

2. Add about 50 mg of fine sand.

3. Gently tap the pipet to pack the adsorbent, about 500 mg or so.

4. Put a bit more sand on top to keep the adsorbent from flying around.

5. Now slowly add the solvent, and let it wet the entire column.

6. Dissolve your sample in a minimum amount of solvent—the least polar solvent in which it'll dissolve—and add this solution slowly to the top of the pipet. Don't fill the pipet with liquid; drip the solution onto the sand.

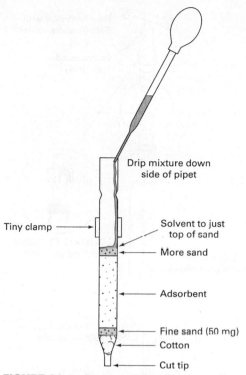

Drip mixture down
side of pipet

Tiny clamp

Solvent to just
top of sand

More sand

Adsorbent

Fine sand (50 mg)

Cotton

Cut tip

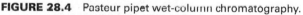

FIGURE 28.4 Pasteur pipet wet-column chromatography.

7. With the sample all on the column, just below the sand add a small amount of **elution solvent**. This may or may not be different from the solvent you dissolved your sample in.

8. Wait and watch. Watch and wait. Do not ever let the column run dry. Collect the drops off the end in a series of fractions in some small containers.

FLASH CHROMATOGRAPHY

Take an ordinary wet-column chromatography setup, put it under pressure, and you've got **flash chromatography**. Usually this takes special apparatus (Fig. 28.5), but the hard part, construction of the column and putting the sample on the column, is as described above. There have been a number of modifications to adapt this technique to the existing, well, less expensive equipment of the undergraduate organic laboratory. These range from using a fish aquarium air pump to provide the necessary air pressure (but I do have concerns about solvents, some of which might be very flammable in the proximity of non-laboratory-approved switches and motors) to placing a Y-shaped glass tube in a single-holed rubber stopper, connecting a source of air or N_2 gas to one end, and "regulating" the pressure to the column with a well-placed thumb on the other arm of the Y.

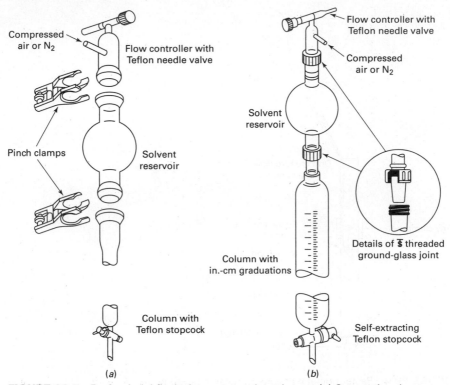

FIGURE 28.5 Professional flash chromatography columns. (*a*) Conventional. (*b*) Screw-thread. From Mayo, Pike, and Forbes, *Microscale Organic Laboratory*, 5th edition, John Wiley & Sons, Inc., New York, 2011.

And because the column is under pressure, the caution of packing the column or jamming the cotton or glass wool into the tip too tightly now has a double edge to it: Go ahead! . . . Packing the column too tightly will matter less, as the gas pressure will compensate and push the solvent through anyway. Hold it! . . . Packing the column too tightly with the whole system under gas pressure might be considered foolhardy. You, unfortunately, must be the judge. And don't forget to release the pressure if you decide to stop the flow at the stopcock or pinch clamp, unless you believe the system can hold up under the pressure.

MICROSCALE FLASH CHROMATOGRAPHY

Put a rubber bulb over the Pasteur pipet wet-column chromatography setup (Fig. 28.4) and press it. Don't let go, unless you want the eluent to slurp back up to the bulb and ruin the setup. Remove the bulb, let it go, put it back on the pipet . . . press. Remove the bulb, let it go, put it back on the pipet . . . press. Remove the bulb, let it go, put it back on the pipet . . . press. Remove the bulb . . .

EXERCISES

1. Compare/contrast wet-column chromatography with TLC.

2. If you don't want the column packing coming out of the column, you have to cram the cotton into the tip with lots of force. Right?

3. Why don't we want cracks or voids in our adsorbents in the column?

REFRACTOMETRY

When light travels from one medium to another, it changes velocity and direction a bit. If you've ever looked at a spoon in a glass of water, you've seen this: The image of the spoon in water is displaced a bit from the image of the spoon in air, and the spoon looks broken. When the light rays travel from the spoon in the water and break out into the air, they are **refracted**, or shifted (Fig. 29.1). If we take the ratio of the sine of the angles formed when

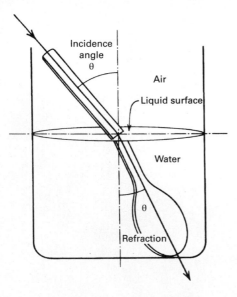

FIGURE 29.1   The refraction of light.

a light ray travels from air to water, we get a single number, the **index of refraction**, or **refractive index**. Because we can measure the index of refraction to a few parts in 10,000, this is a very accurate physical constant for identifying a compound.

The refractive index is usually reported as n_D^{25}, where the tiny 25 is the temperature at which the measurement was taken and the tiny capital D means we've used light from a sodium lamp, specifically a single yellow frequency called the sodium D line. Fortunately, you don't have to use a sodium lamp if you have an Abbé refractometer.

THE ABBÉ REFRACTOMETER

The refractometer looks a bit like a microscope. It has:

1. *An eyepiece.* You look in here to make your adjustments and read the refractive index.

2. *A compensating prism adjust drum.* Since the Abbé refractometer (Fig. 29.2) uses white light and not light of one wavelength (the sodium D line), the white

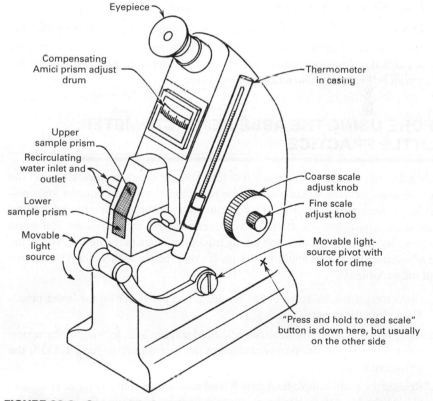

FIGURE 29.2 One model of a refractometer.

light *disperses* as it goes through the optics, and rainbowlike color fringing shows up when you examine your sample. By turning this control, you rotate some Amici compensation prisms that eliminate this effect.

3. *Hinged sample prisms.* This is where you put your sample.

4. *Light source.* This provides light for your sample. It's on a movable arm, so you can swing it out of the way when you place your samples on the prisms.

5. *Light source swivel-arm pivot and lock.* This is a large, slotted nut that works itself loose as you move the light source up and down a few times. Always have a dime handy to help you tighten this locking nut when it gets loose.

6. *Scale adjust knob.* You use this knob to adjust the optics so that you see a split field in the eyepiece. The refractive index scale also moves when you turn this knob. The knob is often a dual control; use the outer knob for a coarse adjustment and the inner knob for a fine adjustment.

7. *Scale/sample field switch.* Press the "Press and hold to read scale" switch, and the numbered refractive index scale appears in the eyepiece. Release this switch, and you see your sample in the eyepiece. Some models don't have this type of switch. You have to change your angle of view (shift your head a bit) to see the field with the refractive index reading.

8. *Line cord on–off switch.* This turns the refractometer light source on and off.

9. *Recirculating water inlet and outlet.* These are often connected to temperature-controlled recirculating water baths. The prisms and your samples in the prisms can all be kept at the temperature of the water.

BEFORE USING THE ABBÉ REFRACTOMETER: A LITTLE PRACTICE

This is going to sound very strange, but before you use the refractometer to get the refractive index of your unknown, use the refractometer on a common known— **water**! See, there are so many things you can vary—lamp position, compensating prisms, scale adjustment—well, if you've never used this refractometer, you could misadjust, say, the lamp position and totally miss the change in the field and drive the scale adjustment past the point of no return. If you practice with water, you remove one of the variables.

1. Open the prisms, make sure they're clean, and put water on the lower prism. Then close the prisms.

2. Now, press and hold the scale/sample field switch, and, looking at the uppermost scale, carefully, purposefully set the instrument to read 1.3333, the refractive index of water.

3. Release the scale/sample field switch, and now start at step 3 in the next section, "Using the Abbé Refractometer," to see the effects of moving the lamp position and the compensating prism adjustment drum.

USING THE ABBÉ REFRACTOMETER

1. Make sure the unit is plugged in. Then turn the on–off switch to ON. The light at the end of the movable arm should come on.

2. Open the hinged sample prisms. *Without touching the prisms at all,* place a few drops of your liquid on the lower prism. Then swing the upper prism back over the lower one and *gently* close the prisms. *Never touch the prisms with any hard object, or you'll scratch them.*

3. Raise the light on the end of the movable arm so that the light illuminates the upper prism. Get out your dime and, with the permission of your instructor, tighten the light-source swivel arm locknut, as it gets tired and lets the light drop.

4. Look in the eyepiece. Slowly, carefully, with *very little* force, turn the large scale and sample image adjust knob from one end of its rotation to the other. *Do not force!* (If your sample is supposedly the same as that of the last person to use the refractometer, you shouldn't have to adjust this much, if at all.)

5. Look for a split optical field of light and dark (Fig. 29.3). This may not be very distinct. You may have to raise or lower the light source and scan the sample a few times.

6. If you see color fringing at the boundary between light and dark (usually red or blue), slowly turn the compensating prism adjust drum until the colors are at a minimum.

7. Use the image adjust knob to center the split optical field right on the cross-hairs.

8. Press and hold the scale/sample field switch. The refractive index scale should appear (Fig. 29.4). Read the uppermost scale, the refractive index, to four decimal

  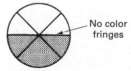

No color fringes

Too high   Too low   Just right

FIGURE 29.3 Your sample through the lens of the refractometer.

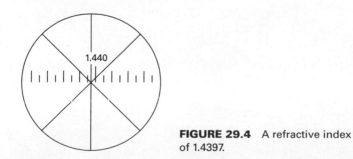

FIGURE 29.4 A refractive index of 1.4397.

places (five digits here). (If your model has two fields, and the refractive index is *always* visible, just read it.)

9. Read and record the temperature of the refractometer. You'll have to correct the reading for any difference between the temperature of your sample and the temperature given in your literature reference. If you don't have that value, correct the reading to 20°C. The correction is 0.0004 units per degree difference above the reference value.

10. Switch the unit to OFF. The swing-arm light should go off.

11. Clean both prisms with ethanol and a Kimwipe. Dispose of waste properly.

REFRACTOMETRY HINTS

1. The refractive index changes with temperature. If your reading is not the same as that of a handbook, check the temperatures before you despair.

2. Volatile samples require quick action. Cyclohexene, for example, has been known to evaporate from the prisms of unthermostatted refractometers more quickly than you can obtain the index. It may take several tries as you readjust the light, turn the scale and sample image adjust, and so on.

3. Make sure the instrument is level. Often, organic liquids can seep out of the prism jaws before you are ready to make your measurement.

Chapter 30

GAS CHROMATOGRAPHY

Gas chromatography (GC) can also be referred to as **vapor-phase chromatography** (VPC) and even **gas-liquid chromatography** (GLC). Usually, the technique, the instrument, and the recording of the data are called **GC**:

"Fire up the GC."	(the instrument)
"Analyze your sample by GC."	(perform the technique)
"Get the data off the GC."	(analyze the chromatogram)

I've mentioned the similarity of all chromatography, and just because electronic instrumentation is used, there's no need to feel that something basically different is going on.

THE MOBILE PHASE: GAS

In **column chromatography**, the mobile (moving) phase is a liquid that carries your material through an adsorbent. I called this phase the **eluent**, remember? Here, a gas is used to push, or *carry,* your vaporized sample, and it's called the **mobile phase**.

The **carrier gas** is usually helium, although you can use nitrogen. You use a **microliter syringe** to inject your sample into this gas stream through an **injection port** and then *onto the column*. If your sample is a mixture, the *compounds separate on the column* and reach the **detector** at different times. As each component hits the detector,

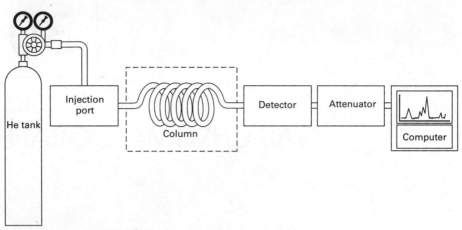

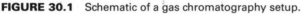

FIGURE 30.1 Schematic of a gas chromatography setup.

the detector generates an **electrical signal**. Usually, the signal goes through an **attenuator network**, and then out to a **computer** to record the signal. I know that's a fairly general description, and Figure 30.1 is a highly simplified diagram. But there are lots of different GCs, so being specific about their operation doesn't help here. You should see your instructor. But that doesn't mean we can't talk about some things.

GC SAMPLE PREPARATION

Sample preparation for GC doesn't require much more work than handing in a sample to be graded. *Clean and dry,* right? Try to take care that the boiling point of the material is low enough that you can actually use this technique. The maximum permissible temperature depends on the type of column, and that should be given. In fact, for any single experiment that uses GC, the nature of the column, the temperature, and most of the electronic settings will be fixed.

GC SAMPLE INTRODUCTION

The sample enters the GC at the **injection port** (Fig. 30.2). You use a microliter syringe to pierce the rubber septum and inject the sample onto the column. Don't stab yourself or anyone else with the needle. Remember, this is *not* dart night at the pub. Don't throw the syringe at the septum. There is a way to do this.

1. To load the sample, put the needle into your liquid sample, and *slowly* pull the plunger to draw it up. If you move too fast and more air than sample gets in, you'll have to push the plunger back again and draw it up once more. Usually, they give you a 10-μL syringe, and 1 or 2 μL of sample is enough. Take the

228

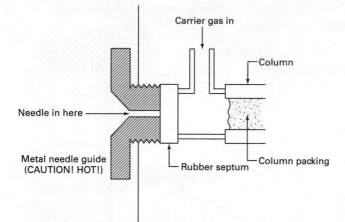

FIGURE 30.2 A GC injection port chromatography setup.

loaded syringe out of the sample, and *carefully, cautiously* pull the plunger back so that there is no sample in the needle. You should see a bit of air at the very top but not very much. This way, you don't run the risk of having your compound boil out of the needle as it enters the injector oven just before you actually inject your sample. That premature injection broadens the sample and reduces the resolution. In addition, the air acts as an internal standard. Since air travels through the column almost as fast as the carrier gas, the **air peak** that you get can signal the start of the chromatogram, much like the pencil line at the start of a TLC plate. Ask your instructor.

2. Hold the syringe in *two hands* (Fig. 30.3). There is no reason to practice being an M.D. in the organic laboratory.

3. Bring the syringe to the level of the injection port, straight on. No angles. Then let the needle touch the septum at the center.

4. The real tricky part is holding the barrel and, without injecting, pushing the needle through the septum. This is easier to write about than it is to do the first time. Pierce the septum carefully.

5. Now quickly and smoothly push on the plunger to inject the sample, and then pull the syringe needle out of the septum and injection port.

After a while, the septum gets full of holes and begins to leak. Usually, you can tell that you have a leaky septum when the trace on the computer wanders about aimlessly without any sample injected.

Speaking of computers, as soon as you do inject that sample, please tell your computer that you have done so. It's been waiting, wagging its tail like a happy puppy, for you to give it the signal to run.

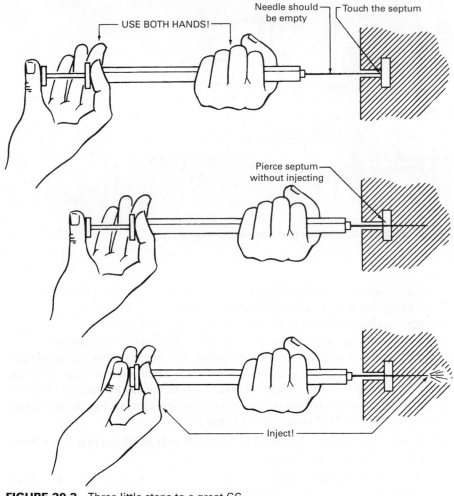

FIGURE 30.3 Three little steps to a great GC.

SAMPLE IN THE COLUMN

Now that the sample is in the column, you might want to know what happens to this mixture. Did I say mixture? Sure. Just as with **thin-layer** and **column chromatography**, you can use GC to determine the composition or purity of your sample.

Let's start with two components, A and B again, and follow their path through an **adsorption column**. Well, if A and B are different, they are going to stick on the adsorbent to different degrees and spend more or less time flying in the carrier gas. Eventually, one will get ahead of the other. Aha! *Separation*—just like column and thin-layer chromatography. Only here the samples are vaporized, and it's called vapor-phase chromatography (VPC).

Some of the adsorbents are coated with a **liquid phase**. Most are very high-boiling liquids, and some look like waxes or solids at room temperature. Still, they're liquid phases. So, the different components of the mixture you've injected will spend different amounts of time in the liquid phase, and, again, you'll get a separation of the components in your sample. Thus, the technique is known as **gas-liquid chromatography (GLC)**. You could use the same adsorbent and different liquid phases and change the characteristics of each column. Can you see how the sample components would partition themselves between the gas and liquid phases and separate according to, perhaps, molecular weight, polarity, size, and so on, making this technique also known as **liquid-partition chromatography**?

Since these liquid phases on the adsorbent are eventually liquids, you can boil them. And that's why there are temperature limits for columns. It is not advisable to heat a column past the recommended temperature, boiling the liquid phase right off the adsorbent and right out of the instrument.

High temperature and air (oxygen) are death for some liquid phases, since they oxidize. So make sure the carrier gas is running through them at all times, even a tiny amount, while the column is hot.

SAMPLE AT THE DETECTOR

There are several types of **detectors**, devices that can tell when a sample is passing by them. They detect the presence of a sample and convert it to an electrical signal that's turned into a **GC peak** (Fig. 30.4) on the computer screen. The most common type is the **thermal conductivity detector**. Sometimes called hot-wire detectors, these devices are very similar to the filaments you find in lightbulbs, and they require some care. Don't ever turn on the **filament current** unless the carrier gas is flowing. A little air (oxygen), a little heat, a little current, and you get a lot of trouble replacing the burned-out detector.

Usually, there are at least two thermal conductivity detectors in the instrument, in a "bridge circuit." Both detectors are set in the gas stream, but only one sees the samples. The electric current running through them heats them up, and they lose heat to the carrier gas at the same rate.

As long as no sample, only carrier gas, goes over *both* detectors, the bridge circuit is **balanced**. There is no signal to the computer, and the trace or value displayed does not move from zero.

Now a sample in the carrier gas goes by one detector. This sample has a thermal conductivity different from that of pure carrier gas. So the **sample detector** loses heat at a different rate from the **reference detector**. (Remember, the **reference** is the detector that NEVER sees samples—only carrier gas.) The detectors are in different surroundings. They are not really equal anymore. So the bridge circuit becomes unbalanced, and a signal goes to the computer, giving a GC peak.

Try to remember the pairing of **sample** with **reference**; it's the *difference* between the two that most electronic instrumentation responds to. You will see this again and again.

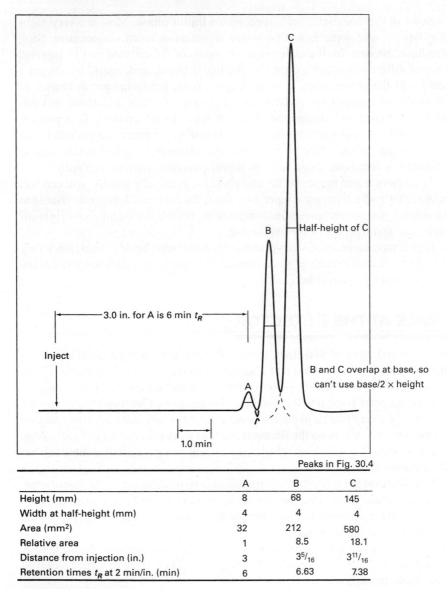

FIGURE 30.4   A well-behaved GC trace of a mixture of three compounds.

Peaks in Fig. 30.4

	A	B	C
Height (mm)	8	68	145
Width at half-height (mm)	4	4	4
Area (mm²)	32	212	580
Relative area	1	8.5	18.1
Distance from injection (in.)	3	$3^5/_{16}$	$3^{11}/_{16}$
Retention times t_R at 2 min/in. (min)	6	6.63	7.38

ELECTRONIC INTERLUDE

The electrical signal makes two other stops on its way to the computer.

1. ***The coarse attenuator.*** This control makes the signal weaker (attenuates it). Usually, there's a scale marked in **powers of two**: 2, 4, 8, 16, 32, 64, . . ., so each position is half as sensitive as the last one. There is one setting, either an

∞ or an **S** (for **shorted**), which means that the attenuator has *shorted out the terminals connected to the computer.* Now the **computer software zero** can be set properly.

2. *The GC zero control.* This control helps set the zero position on the computer, but it is *not to be confused with the zero control in the computer software.*

Here's how to set up the electronics properly for a GC and a data-collection program.

1. The computer should be booted up, and the GC data collection software started. The **GC** should be allowed to warm up and stabilize for at least 10–15 minutes. Some systems take more time; ask your instructor.

2. Set the **coarse attenuator** to the highest attenuation, usually an ∞ or an S. With the GC output shorted, no signal reaches the computer.

3. Now, set the computer readout—either a trace or a voltage value—to zero, using *only the computer software zero control.* Once you do that, *leave it alone.* This way, you're not introducing a zero-offset error—a reading either added to or subtracted from the signal from the GC by the computer.

4. Start turning the *coarse attenuator* control to more sensitive settings (lower numbers), and *watch the trace on the computer.*

5. If the computer reading moves off zero, *use only the GC zero control* to bring the trace back to the zero point on the computer screen. *Do not use the computer software to get a reading of zero. Use only the GC zero control.*

6. As the *coarse attenuator* reaches more sensitive settings (lower numbers), it becomes more difficult to adjust the *trace to zero* using *only the GC zero control.* Do the best you can at the *lowest attenuation* (highest sensitivity) you can hold a zero steady at.

7. Now, you *don't* normally run samples on the GC at attenuations of 1 or 2. These settings are *very sensitive,* and there may be lots of electrical noise—the trace jumps about. The point is, *if the GC zero is OK at an attenuation of 1,* then when you run at attenuations of 8, 16, 32, and so on, *the baseline will not jump if you change attenuation in the middle of the run.*

Now that the attenuator is set to give peaks of the proper height, you're ready to go. Just be aware that there may be a **polarity switch** that can make your peaks shift direction.

SAMPLE ON THE COMPUTER

Basic computer analysis of a GC trace can be pretty simple:

1. *Mark your starting point.* This is the same as that mark you put on TLC plates to show where you spotted your sample. If you have an air peak, set your starting point there. If you don't have an air peak, the starting point would have to be when you first had the computer start collecting data. Can you set

the starting point anywhere else? Some software packages let you set the start anywhere during a run. For the basic analysis of getting the retention time, I'm not sure how useful that would be.

2. *Mark the peaks.* Use the software to give the retention times. Usually you select the top of a peak, give a mouse click, and behold—the retention time. If you used an air peak, you can get the software to say that the air peak is at time = 0.0, and go from there.

3. *Get the areas.* Sometimes, right after you have identified the peaks you want by selecting them, the software is ready to print out the relative peak areas right away. In a GC trace with a good, stable baseline, this is probably all you need. If you have baseline drift, or ill-resolved (partially overlapping) peaks, the simple, automatically calculated values might not do the job.

4. *Massage the areas.* Again, you might not have to do this with a well-behaved GC trace. Otherwise, you'll have to use special baseline-correction functions, or even peak-separation functions, in the software to get more accurate results. For baseline correction, you might have to draw or click a baseline under a particular peak yourself before letting the computer calculate the area under the peak.

With peak areas you can calculate how much of each compound is in your sample.

PARAMETERS, PARAMETERS

To get the best GC trace from a given column, there are lots of things you can do, simply because you have so many controls. Usually you'll be told the correct conditions, or they'll be preset on the GC.

Gas Flow Rate

The faster the carrier gas flows, the faster the compounds are pushed through the column. Because they spend *less time in the stationary phase,* they don't separate as well, and the *GC peaks come out very sharp but not well separated.* If you *slow the carrier gas down too much,* the compounds spend so much time in the stationary phase that *the peaks broaden and the overlap gets very bad.* The optimum is, as always, the best separation you need, in the shortest amount of time. Sometimes you're on your own. Most of the time, someone else has already worked it out for you.

Temperature

Whether or not you realize it, the GC column has its own heater—the **column oven**. If you turn the temperature up, the compounds hotfoot it through the column very quickly. Because they spend *less time in the stationary phase,* they don't separate as well, and the *GC peaks come out very sharp but not well separated.* If you *turn* *the temperature down some,* the compounds spend so much time in the stationary phase that *the peaks broaden and overlap gets very bad.* The optimum is, as always, the best separation you need in the shortest amount of time. There are two absolute limits, however.

1. *Too high a temperature, and you destroy the column.* The adsorbent may decompose, or the liquid phase may boil out onto the detector. *Never exceed the recommended maximum temperature for the column material.* Don't even come within 20°C of it.

2. *Way too low a temperature, and the material condenses on the column.* The temperature has to be above the **dew point**, not the boiling point, of the least volatile material. Water doesn't always condense on the grass—*become dew*—every day that's just below 100°C (that's 212°F, the boiling point of water). Fortunately, you don't have to know the dew points for your compounds. However, you *do* have to know that the temperature doesn't *have to be above the boiling point* of your compounds.

Incidentally, the **injector** may have a separate **injector oven**, and the **detector** may have a separate **detector oven**. Set them both 10–20°C higher than the column temperature. You can even set these *above the boiling points* of your compounds, since you *do not want them to condense in the injection port or the detector, ever.* For those with *only one temperature control,* sorry. The injection port, column, and detector are all in the same place, all in the same oven, and all at the same temperature. The *maximum temperature,* then, is *limited by the decomposition temperature of the column.* Fortunately, because of that dew point phenomenon, you really don't have to work at the boiling points of the compounds.

EXERCISES

1. How do the temperature and the flow rate of the carrier gas affect the retention time of two separable compounds? Does the amount of separation also change? If so, how?

2. If air can degrade a column and cause TC detector filaments to burn out, why do we use an air peak?

HP LIQUID CHROMATOGRAPHY

HPLC. Is it **high-performance liquid chromatography** or **high-pressure liquid chromatography**, or something else? It's probably easier to consider it a delicate blend of **wet-column chromatography** and **gas chromatography** (see Chapters 28 and 30, respectively).

Rather than letting gravity **pull** the solvent through the powdered adsorbent, the liquid is **pumped** through under pressure. Initially, high pressures [1000–5000 psig (**pounds per square inch gauge**), that is, not absolute] were used to push a liquid through a tightly packed solid. But the technique works well at lower pressures (250 psig); hence the name high-performance liquid chromatography.

From there on, the setup (Fig. 31.1) closely resembles gas chromatography. Although a **moving liquid phase** replaces the helium stream, compounds are put onto a column either manually, through an **injection port**, or using an **automatic injector**. They *separate inside a chromatographic column* in the same way as in GC by spending more or less time in a *moving liquid,* and the separated compounds pass *through a detector*. There, as the amounts of each compound go by the **detector**, they are turned into electrical signals and *displayed on a computer screen as HPLC peaks* that look just like GC peaks. You should also get the feeling that the analysis of these HPLC traces is done similarly to GC traces, because it is.

Again, there's a lot of variety among HPLC systems, so what I say won't necessarily apply to your own system in every respect. But it should help.

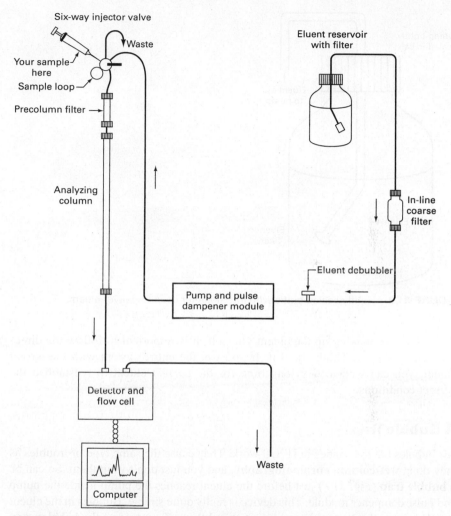

Six-way injector valve

Waste

Your sample here

Sample loop

Precolumn filter →

Analyzing column

Eluent reservoir with filter

In-line coarse filter

Eluent debubbler

Pump and pulse dampener module

Detector and flow cell

Computer

Waste

FIGURE 31.1 Block diagram of an HPLC setup with manual injection.

THE MOBILE PHASE: LIQUID

If you use only *one liquid,* either **neat** or as a mixture, the entire chromatogram is said to be **isochratic**. Units that can deliver *varying solvent compositions over time* are called **gradient elution systems**.

For an **isochratic** system, you usually use a single solution, or a neat liquid, and put it into the **solvent reservoir**, which is generally a glass bottle that has a tube going through the cap, to a filter that you put into the eluent (Fig. 31.2). The solvent travels out of the reservoir and into the **pump and pulse dampener module**.

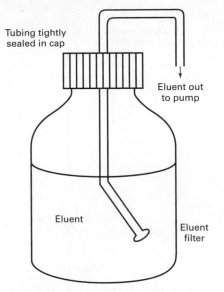

FIGURE 31.2 Tightly sealed bottle with internal filter used to deliver eluent.

If you're making up the eluent yourself, it is important to follow the directions *scrupulously*. Think about it. If you wet the entire system with the wrong eluent, you can wait a very long time for the *correct* eluent to reestablish the correct conditions.

A Bubble Trap

Air bubbles are the nasties in HPLC work. They cause the same type of troubles as they do in **wet-column chromatography**, and you just don't want them. So can be a **bubble trap** (Fig. 31.3) just before the eluent reaches the pump inside the pump and pulse dampener module. This device is really quite simple. Bubbles in the eluent stream rise up the center pipe and are trapped there. To get rid of the bubbles, you open the cap at the top. Then, solvent rises in the tube and pushes the bubbles out. You have to be *extremely* careful about bubbles if you're the one to start the setup or if the solvent tank has run out. Normally, *one bubble purge per day* is enough.

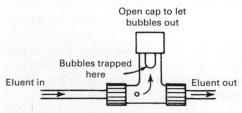

FIGURE 31.3 Common bubble trap.

The Pump and Pulse Dampener Module

The most common pumping system is the **reciprocating pump**. The pump has a **reciprocating ruby rod** that moves back and forth. On the backstroke, the pump loads up on a little bit of solvent; then, on the forward stroke, it squirts it out under pressure. If you want to increase the amount of liquid going through the system, you can have the computer change the length of the stroke from zero to a preset maximum. The software and controller shouldn't let anyone start the pump at maximum; a good smack might damage the piston.

Because the rod **reciprocates** (i.e., goes back and forth), you'd expect huge swings in pressure, and pulses of pressure do occur. That's why they make **pulse dampeners**. A coiled tube is hooked to the pump on the side opposite the column. It is filled with the eluent that's going through the system. On the forward stroke, solvent is compressed into this tube, and simultaneously a shot of solvent is pushed onto the column. On the backstroke, while the pump chamber fills up again, the eluent we just pressed into the pulse dampener *squirts out into the column*. Valves in the pump take care of directing the flow. As the eluent in the pulse dampener tubing takes up the slack, the huge variations in pressure, from essentially zero to perhaps 1000 psig, are evened out. They don't disappear, going about 100 psig either way, but these **dampened pulses** are now too small to be picked up on the detector. They don't show up on the computer screen either.

HPLC SAMPLE PREPARATION

Samples for HPLC must be *liquids* or *solutions*. It would be nice if the solvent in which you've dissolved your solid sample were the same as the eluent.

It is *absolutely crucial* that you preclean your sample. *Any* decomposed or insoluble material will stick to the top of the column and can continually poison further runs. There are a few ways to keep your column clean.

1. *The Swinney adapter* (Fig. 31.4). This handy unit locks onto a syringe *already* filled with your sample. Then you push the sample *slowly* through a Millipore filter to trap insoluble particles. This does *not*, however, get rid of **soluble tars** that can ruin the column. (Oh: Don't confuse the Millipore filters with the papers that separate them. It's embarrassing.)

2. *The precolumn filter* or *guard column* (Fig. 31.5). Add a tiny column filled with exactly the same material as the main column, and *let this small column get contaminated*. After a while, it used to be that you removed it, cleaned out the gunk and adsorbent, and refilled it with fresh column packing. Today it just gets thrown out. Either way, you don't really know when the garbage is going to poison the entire precolumn filter and then start ruining the analyzing column. The only way to find out whether you have to clean the precolumn is to take it out of the instrument. You really want to clean it out or throw it out long before the contaminants start to show up at the precolumn, or guard column, exit.

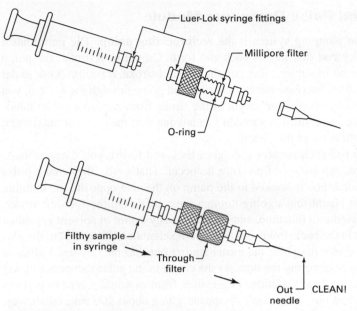

FIGURE 31.4 The Swinney adapter and syringe parts. (Note Luer-Lok.)

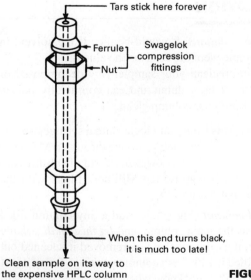

FIGURE 31.5 The precolumn filter.

HPLC SAMPLE INTRODUCTION

This uses the equivalent of the **injection port** for the GC technique. With GC, you could inject through a rubber septum directly onto the column. With HPLC, it's very difficult to inject against a liquid stream moving at possibly 1000 psig. That's why **injection port valves** were invented for HPLC: You put your sample into an **injection loop** on the valve that is *not in the liquid stream,* then *turn the valve,* and *voilà, your sample is in the stream,* headed for the column.

If you have an **autosampler**, this same valve exists and works the same way (so you should read the rest of this section), with a computer-controlled motor to flip the valve from one position to the other. And instead of a manual syringe, there's a computer-controlled syringe that injects the samples into the valve.

The injection port valve (Fig. 31.6) has two positions:

1. *Normal solvent flow.* In this position, the eluent enters the valve, goes around, and comes on out into the column without any bother. *You put the sample in the injection loop in this position.*

2. *Sample introduction.* Flipped this way, the valve allows the eluent to be *pumped through the injection loop*, and any sample there is carried along and into the column. *You put the sample on the column in this position.*

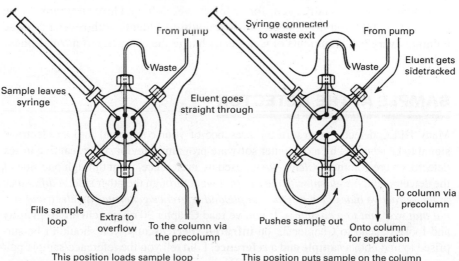

FIGURE 31.6 The sample injector valve.

An HPLC autosampler, besides having that automated injection loop valve, needs motors to move a sample tray, and the sample pickup tubing, and air to move the sample into and out of the tubing, and its own solvent(s) to pre-wash and post-wash the sample pickup tubing. Phew!

If all of that is taken care of for you, and all you have to do is to put your properly prepared sample into the rack, just make sure yours is in the right position.

SAMPLE IN THE COLUMN

Once the sample is in the column, there's not much difference between what happens here and what happens in paper, thin-layer, vapor-phase (gas), or wet-column chromatography. *The components in the mixture will remain in the stationary phase or move in the mobile phase for different times and end up at different places when you stop the experiment.*

So what's the advantage of HPLC? You can separate and detect microgram quantities of solid samples much as in GC, but for GC your solid samples have to vaporize without decomposing. Fat chance of that. You can analyze a solid sample without decomposing it with HPLC.

HPLC also uses something called a **bonded reversed-phase column**, in which the stationary phase is a nonpolar hydrocarbon that is chemically bonded to a solid support. You can use these with aqueous eluents, usually alcohol–water mixtures. So you have a *polar eluent and a nonpolar stationary phase*, something that does not usually occur for ordinary wet-column chromatography. One advantage is that you don't need to use anhydrous eluents with reversed-phase columns (very small amounts of water can change the character of normal-phase columns).

SAMPLE AT THE DETECTOR

Many HPLC detectors can turn the presence of your compound into an electrical signal to be acquired by a computer software program. Previously, **refractive index** detectors were common. *Clean eluent*, used as a reference, went through one side of the detector, and the *eluent with the samples* went through the other side. *A difference in the refractive index between the sample and reference generated an electrical signal that was sent to a computer.* If you've read Chapter 30 on **gas chromatography** and looked ahead to Chapter 32 on **infrared spectroscopy**, you shouldn't be surprised to find both a **sample** and a **reference**. I did tell you the reference/sample pair is common in instrumentation (Fig. 31.7).

Ultraviolet (UV) detectors are more sensitive than refractive index detectors, and, if you wanted to do gradient elution—where two or more solvents mixed in changing compositions make up the eluent—you'd be out of luck. Continuously splitting out a clean, variable composition reference solvent is very difficult.

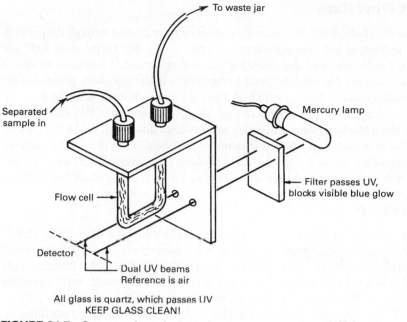

Separated sample in

To waste jar

Mercury lamp

Filter passes UV, blocks visible blue glow

Flow cell

Detector

Dual UV beams
Reference is air

All glass is quartz, which passes UV
KEEP GLASS CLEAN!

FIGURE 31.7   Cutaway view of a very simple sample/reference HPLC UV detector.

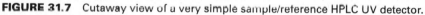

But then again, eluents used with UV detectors shouldn't absorb a lot of UV at wavelengths that you might be looking for your samples to absorb at.

And in certain cases, you can't really use the area under the HPLC curve to tell you how much of each component is in your sample if the components have differing UV absorptions at your analytical wavelength. Since many HPLC setups use UV diode array detectors, you might want to look at the UV-VIS section in Chapter 32.

SAMPLE ON THE COMPUTER

Go back and read about HPLC peak interpretation in the sections on GC detector electronic setup ("Electronic Interlude" in Chapter 30) and peak interpretation ("Sample on the Computer" in Chapter 30). The analysis is *exactly the same*—retention times, peak areas, baselines . . . all that.

PARAMETERS, PARAMETERS

To get the best HPLC trace from a given column, there are lots of things you can do; most of them are the same as those for GC (see "Parameters, Parameters" in Chapter 30).

Eluent Flow Rate

The faster the eluent flows, the faster the compounds are pushed through the column. Because they spend *less time in the stationary phase,* they don't separate as well, and the *HPLC peaks come out very sharp but not well separated.* If you *slow the eluent down too much,* the compounds spend so much time in the stationary phase that the *peaks broaden and overlap gets very bad.* The optimum is, as always, the best separation you need in the shortest amount of time. One big difference in HPLC is the need to worry about **back-pressure**. If you try for very high flow rates, the HPLC column packing tends to collapse under the pressure of the liquid. This, then, is the cause of the back-pressure: *resistance of the column packing.* If the pressures get *too high,* you may burst the tubing in the system, damaging the pump . . . all sorts of fun things.

Temperature

Not many HPLC setups have ovens for temperatures like those for GC because *eluents tend to boil at temperatures much lower than the compounds on the column,* which are usually solids anyway. Eluent bubbling problems are bad enough without actually boiling the solvent in the column. This is not to say that HPLC results are independent of temperature. They're not. But if a column oven for HPLC is present, its purpose more likely is to keep stray drafts and sudden chills away than to have a hot time.

Eluent Composition

You can vary the composition of the eluent (mobile phase) a lot more in HPLC than in GC, so there's not really much correspondence. Substitute nitrogen for helium in GC, and usually the sensitivity decreases, but the retention times stay the same. Changing the mobile phases—the gases—in GC doesn't have a very big effect on the separation or retention time.

There are much better parallels to HPLC: **TLC** or **column chromatography**. Vary the eluents in these techniques, and you get widely different results. With a **normal-phase, silica-based column**, you can get results similar to those from silica gel TLC plates.

EXERCISES

1. Compare and contrast HPLC with GC.

2. Gradient elution, where the solvent changes as the HPLC is run (much like temperature programming in GC), is useful in what kinds of cases? Someone programmed the HPLC to start with the most polar solvent and then switch over to a less polar one, the reverse of the usual case. Is this an error, or are there applications where this might be useful?

3. You've just finished running an HPLC using a single wavelength in your UV detector, and your HPLC trace has two peaks of equal area. Can you say the initial composition was 50:50? What if one component was only 10%, but had an absorption ten times higher than the other component at that analytical wavelength?

INFRARED SPECTROSCOPY

(And A Bit Of UV-VIS, Too)

In **infrared (IR) absorption spectroscopy**, infrared radiation of wavelengths from 2.5 to 15 micrometers (μm; microns) gets passed through (usually) or reflected from your product, and, as the instrument scans your sample, *specific functional groups absorb at specific energies*. Because this spectrum of energies is laid out on a piece of paper, these specific energies become specific places on a computer screen or chart.

MOLECULES AS BALLS ON SPRINGS

You can model a diatomic molecule as two balls of mass m connected by a spring with a force constant (stiffness) k. This is called the **classical harmonic oscillator** that we know and love from physics. At equilibrium distance (d), the displacement from this position (x) is zero, and the spring has no potential energy. Stretch the spring to $+x$, or compress the spring to $-x$, and the energy increases along the curve. That curve is a parabola called the **harmonic potential** (Fig. 32.1).

If you get this system going from $+x$ through zero to $-x$ and back, again and again, you've set up a vibration. And the frequency (v) of that vibration is given by the formula

$$v = \frac{1}{2\pi} \sqrt{\frac{k}{m}} \qquad (1)$$

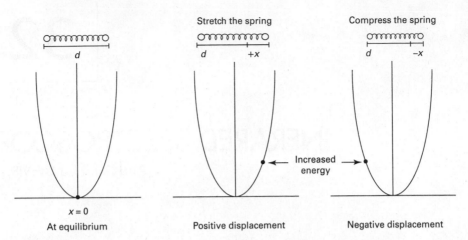

FIGURE 32.1 Harmonic potential of a simple ball-and-spring oscillator.

Strictly speaking, for a system of two masses connected by a spring, m here is replaced by the reduced mass μ (mu):

$$\mu = \frac{m_1 \cdot m_2}{m_1 + m_2} \tag{2}$$

Nonetheless, equation (1) shows that if you have either a stronger spring (larger force constant, k) or a smaller mass (m), then the frequency of oscillation is going to go up. A weaker spring or a larger mass, and the frequency of the vibration will go down.

This ball-and-spring model will vibrate with a total energy (E) equal to the potential energy at the stretched or compressed position (Fig. 32.2):

$$E = \frac{1}{2} k x_{max\ or\ min}^2 \tag{3}$$

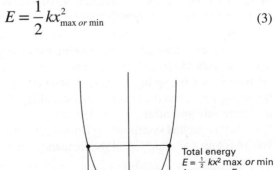

Total energy
$E = \frac{1}{2} k x^2$ max *or* min
Any x; any E

FIGURE 32.2 An unrestricted position for x gives any E.

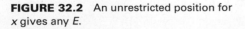

246

While balls on springs can vibrate at any energy (and any frequency), molecules, even simple diatomic ones, don't have this luxury.

AH, QUANTUM MECHANICS

Quantum mechanics predicts that the molecule may vibrate only at energy levels that fit the formula

$$E_n = \left(n + \frac{1}{2} \right) hv \qquad\qquad n = 0, 1, 2, 3 \qquad (4)$$

where v is the frequency of vibration and h is the Planck's constant (6.6262×10^{-34} joule-second). Because of the limitation to discrete energy levels, the energy is said to be **quantized**. The molecule can only absorb (or emit) light of energy equal to the spacing between two levels, and you can only go from one level to the next (the **selection rule** here is $\Delta n = \pm 1$) (Fig. 32.3).

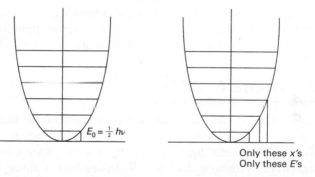

$$E_0 = \tfrac{1}{2}\, hv$$

Only these x's
Only these E's

FIGURE 32.3   Quantization of energy levels sets restrictions.

For the vibrational levels for ordinary molecules, the necessary and sufficient energy to effect a transition from one level to the next is contained in light energy whose frequencies are in the range of the infrared, approximately 4000 to 650 wavenumbers (wavelengths of about 2.5 to 16 μm). More on this later.

THE DISSONANT OSCILLATOR

Unfortunately, the absorptions of real diatomic molecules are much more complicated because they don't behave as harmonic oscillators. Pull the atoms apart too much, and you break the spring—er, bond. Push them together, and they start to repel each other; the energy of the system goes way up very rapidly. This behavior

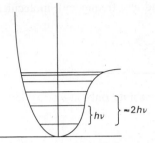

FIGURE 32.4 An anharmonic oscillator allows overtones as well as fundamental vibration.

can be modeled with an **anharmonic potential**, and the vibrating system is usually called an **anharmonic oscillator**. The energy levels aren't equally spaced anymore; you can get transitions of more than one level now, and you can get more than one absorption in the infrared energy spectrum. The transition from $n = 0$ to $n = 1$ is called the **fundamental absorption**, just as before. Now that you can also get transitions from $n = 0$ to $n = 2$ (a different **selection rule**; $\Delta n = \pm 2$), you can get **overtone absorptions** that are almost—not quite—twice the frequency of the fundamental absorption. Why not exactly twice? For the anharmonic oscillator, the energy levels are no longer evenly spaced, so the energies for the transition from $n = 0$ to $n = 1$ won't be the same as those for the transition from $n = 1$ to $n = 2$ (Fig. 32.4).

BUT WAIT! THERE'S MORE

Even with all this, if there's no change in the **dipole moment**, the molecule won't absorb light. If the two masses on the ends of your spring (bond) are hydrogen atoms, and H_2 always has a net dipole of zero, there are no absorptions of light in the infrared spectrum. Hydrogen chloride (HCl) does have a dipole change as it stretches, and, as long as the frequency of the light is correct, will absorb infrared energy. The intensity of that absorption is related directly to the magnitude of that dipole change.

MORE COMPLICATED MOLECULES

The water molecule is apparently not that much more complicated than a diatomic molecule, having only one more atom, but an analysis of the infrared absorption spectrum requires some care. With three atoms and two bonds, a water molecule can appear to be moving around more than those Spandex-clad folk on an exercise video.

The vibrations can be broken down into three simpler motions called **normal modes of vibration**. For water, there are the bend, the symmetric stretch, and the asymmetric stretch (often called the antisymmetric stretch). Every nonlinear molecule has $3N - 6$ normal modes of vibration, where N is the number of atoms in the molecule. Since each normal mode in the water molecule changes the dipole moment as it vibrates, each will absorb infrared light. Three peaks should appear in the infrared spectrum of water, with the two stretching vibrations overlapped because they are similar movements (Fig. 32.5).

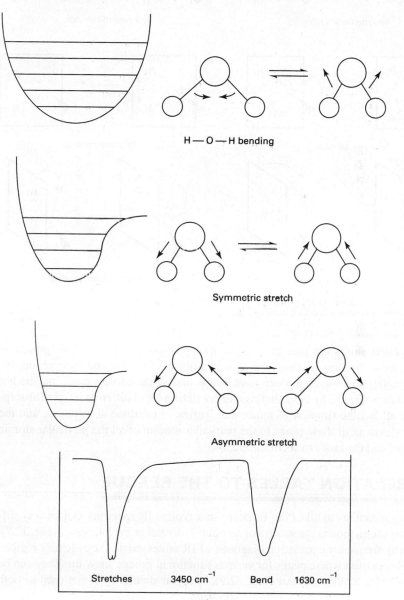

H — O — H bending

Symmetric stretch

Asymmetric stretch

Stretches   3450 cm^{-1} Bend   1630 cm^{-1}

Approximate absorptions of H$_2$O

FIGURE 32.5 Stretches and a bend give rise to the infrared absorption spectrum of water.

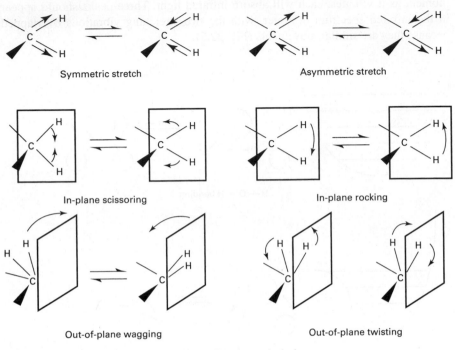

FIGURE 32.6 Some —CH_2— groups going through their paces.

If three atoms are part of a larger molecule—say, the —CH_2— groups in hexane—other bending motions can occur: in-plane scissoring and rocking, out-of-plane twisting and wagging, deep knee bends and squat thrusts—well, maybe not those last two (Fig. 32.6). Nonetheless, there will be a lot of different infrared absorptions for all but the simplest of molecules. Throw in overtone absorptions, and the direct assignment of these peaks to the particular motion of all the particular atomic groupings, and the task can seem formidable.

CORRELATION TABLES TO THE RESCUE

Ascribing meanings to all of the IR peaks in a typical IR spectrum can be very difficult. Just take a look at the spectra of *tert*-butyl alcohol or cyclohexanol (Fig. 32.7). That's why frequency correlation diagrams of IR tables exist. They identify regions of the IR spectrum where peaks for various functional groups show up. They can be graphical (Fig. 32.8) or tabular (Fig. 32.9), and you should try to get used to both styles when interpreting your infrared spectrum.

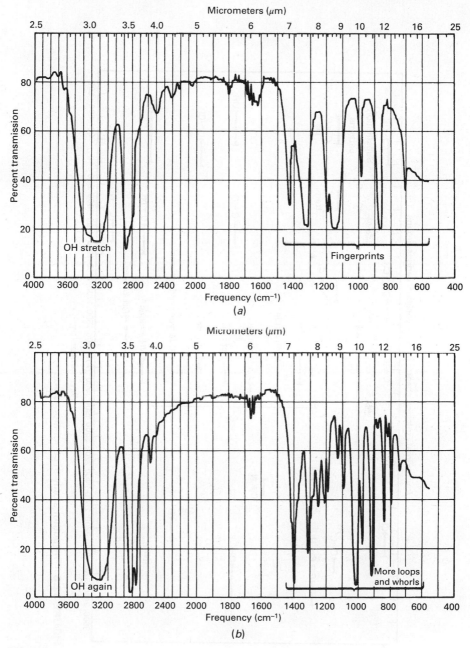

FIGURE 32.7 IR spectra of (*a*) *tert*-butyl alcohol and (*b*) cyclohexanol.

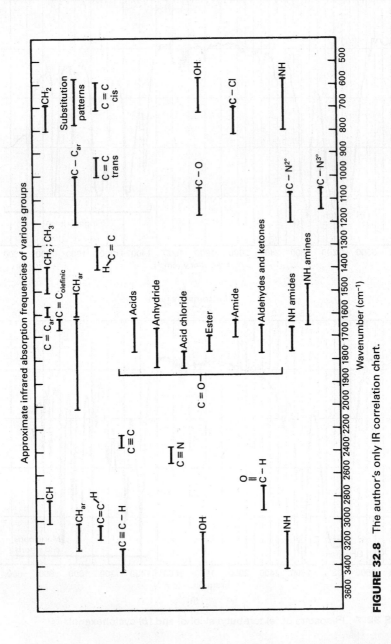

FIGURE 32.8   The author's only IR correlation chart.

Characteristic Infrared Absorptions of Functional Groups

Group	Frequency Range (cm⁻¹)	Intensity[a]
A. Alkyl		
C–H (stretching)	2853–2962	(m–s)
Isopropyl, –CH(CH₃)₂	1380–1385 and 1365–1370	(s)
tert-Butyl, –C(CH₃)₃	1385–1395 and ≈1365	(s)
B. Alkenyl		
C–H (stretching)	3010–3095	(m)
C=C (stretching)	1620–1680	(v)
R–CH=CH₂	985–1000 and 905–920	(s)
R₂C=CH₂	880–900	(s)
cis-RCH=CHR	675–730	(s)
trans-RCH=CHR	960–975	(s)
C. Alkynyl		
≡C–H (stretching)	≈3300	(s)
C≡C (stretching)	2100–2260	(v)
D. Aromatic		
Ar–H (stretching)	≈3030	(v)
Aromatic substitution type (C–H out-of-plane bendings)		
Monosubstituted	690–710 and 730–770	(very s)
ortho-Disubstituted	735–770	(s)
meta-Disubstituted	680–725 and 750–810	(very s)
para-Disubstituted	800–840	(s)
E. Alcohols, Phenols, Carboxylic Acids		
OH (alcohols, phenols, dilute solution)	3590–3650	(sharp, v)
OH (alcohols, phenols, hydrogen bonded)	3200–3550	(broad, s)
OH (carboxylic acids, hydrogen bonded)	2500–3000	(broad, v)
F. Aldehydes, Ketones, Esters and Carboxylic Acids		
C=O stretch (generic)	1630–1780	(s)
aldehydes	1690–1740	(s)
ketones	1680–1750	(s)
esters	1735–1750	(s)
carboxylic acids	1710–1780	(s)
amides	1630–1690	(s)
G. Amines		
N–H	3300–3500	(m)
H. Nitriles		
C≡N	2220–2260	(m)

[a]Abbreviations: s = strong m = medium w = weak v = variable ≈ = approximately

FIGURE 32.9 A tabular IR correlation table. From T. W. G. Solomons and C. B. Fryhle, *Organic Chemistry*, 8th edition, John Wiley & Sons, New York, 2004.

TROUGHS AND RECIPROCAL CENTIMETERS

Although the technique is called infrared absorption spectroscopy, you'll notice that the abscissa (y-axis) is specified in **percent transmission** (as in Figure 32.7) or **%T**. Although you can convert from one scale to the other,

$$Absorbance = 2 - \log_{10}\%T,$$

the absorbance scale, going from 0 to ∞, can cause problems, especially for high values of absorption. Thus the vastly more well-behaved %T scale is used, and thus the "troughs." We still call them peaks, though. (As an exercise, get out your calculator or spreadsheet and solve for the absorbance with %T being zero, or a complete block of radiation.)

Although wavelength (λ) in **micrometers (μm; microns)** can be used as the ordinate (x-axis) units to denote the position of peaks on the infrared spectrum, almost all assignments are given in something called **reciprocal centimeters**, or **wavenumbers (cm^{-1})**. Again, it's a simple calculation to convert from one into the other:

$$cm^{-1} = \frac{1}{\mu m} \times 10000 \qquad \text{and} \qquad \mu m = \frac{1}{cm^{-1}} \times 10000$$

If you look a little more carefully, you'll see that one scale—usually the wavenumber scale—is linear, and the other scale is not. It turns out that absorbances for most **functional groups** are generally in the first half of the spectrum (4000 to 1400 cm^{-1} or 2.5 to 7.2 μm), while the **fingerprint region** ranges from 1400 to 990 cm^{-1} or 7.2 to 11.1 μm. If you plot your spectrum **linear to wavenumbers**, then the first part, *the functional group region, gets expanded,* and the fingerprint region gets compressed. Plot your spectrum **linear to wavelength**, and the front part, *the functional group region, gets compressed* while the fingerprint region gets expanded. Since the purpose of taking the infrared absorption spectrum of your compound is to identify the functional groups present, plots linear to wavenumbers will probably be more useful.

SOME FUNCTIONAL GROUP ANALYSIS

Look at Figure 32.7. Here's a fine example of a pair of alcohols if ever there was one. See the peak (some might call it a trough) at about 3400 cm^{-1} (2.9 μm)? That's due to the **OH group**, specifically the stretch in the O—H bond, the **OH stretch**.

Now consider a couple of ketones, 2-butanone and cyclohexanone (Fig. 32.10). There's no OH peak at about 3400 cm^{-1} (2.9 μm), is there? Should there be? *Of course not.* Is there an OH in 2-butanone? *Of course not.* But there is a C=O, and where's that? The peak is at about 1700 cm^{-1} (5.9 μm). It's *not there* for the alcohols, and it *is there* for the ketones. Right. You've just correlated or *interpreted* four IRs.

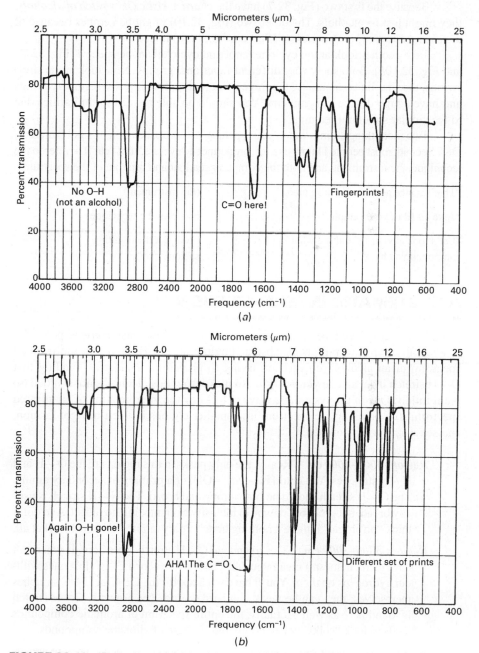

FIGURE 32.10 IR spectra of (a) 2-butanone and (b) cyclohexone.

Because the first two (Fig. 32.7) have the *characteristic OH stretch of alcohols,* they might just be alcohols. The other two (Fig. 32.10) might be ketones because of the *characteristic* C=O *stretch* at 1700 cm^{-1} (5.9 μm) in each.

Take another look at the cyclohexanol and cyclohexanone spectra (Fig. 32.7*b* and Fig. 32.10*b*). Both show very different functional groups. Now look at the similarities, the simplicity, including the fingerprint region. Both are six-membered rings and have a high degree of symmetry. You should be able to see the similarities due to the similar structural features.

Two more things. First, watch your spelling and pronunciation; it's not "infared," OK? Second, most people I know use IR (pronounced "eye-are," not "ear") to refer to the technique, the instrument, and both the computer readout and printout of the spectrum:

"That's a nice new IR you have there."   (the instrument)

"Take an IR of your sample."   (perform the technique)

"Let's look at your IR and see what kind of   (interpret the resulting
compound you have."   spectrogram)

A SYSTEMATIC INTERPRETATION

To interpret an infrared spectrum, you'll need to see if the characteristic peaks for functional groups are present or absent. Finding an O—H stretch in the infrared spectrum of an alcohol you've been trying to prepare is wonderful. Finding that O—H stretch in an alkene you're trying to make from that alcohol is not so good, to put it mildly. One of the systematic ways of tackling the interpretation of an infrared spectrum is to break up the spectrum into smaller regions in a left-to-right fashion, and see what is or is not there.

Region: 4000–3000 cm^{-1}. Really, the only peaks in this region are the O—H and N—H stretches. You could have an —OH, or a primary or secondary amine. The O—H stretch should be both strong and broad. The N—H stretch is usually broad, but probably ranges in intensity from medium to pitiful. And don't think "no amine" if there's nothing here; tertiary amines have no N—H at all.

Region: 3300–3000 cm^{-1}. The *sp* and *sp*^{2} C—H stretches are here. If you don't have them, it doesn't mean you don't have a C=C or a C≡C, just ones without hydrogens on them. You'll have to check for the C=C or C≡C stretches themselves; they can be weak and hard to find. The aromatic C—H stretch is found here as well. Actually, the entire spectrum of aromatic compounds tends to look "spiky" when compared to those of aliphatic compounds.

Region: 3000–2700 cm^{-1}. The *sp*^{3} C—H stretches are here. They're generally fairly sharp, well-defined, yet have overlapping peaks. If these C—H stretches appear to be on top of, or seem to disappear in, a very broad "swelling," check for a C=O peak farther down the line. That's the extremely hydrogen-bonded —OH of a carboxylic acid.

Region: 2250–2100 cm⁻¹. The C≡C and C≡N stretches are here. These tend to be medium-to-weak peaks, with C≡N being the strongest, terminal C≡C—H next, and finally internal C≡C.

Region: 1850–1600 cm⁻¹. This is where C=O stretches are, and they will be the strongest of absorptions. You'll have to do a little more sifting:

- A stretch around 1830–1800 cm⁻¹? You might have an acid chloride or, less likely, an acid bromide. If there is a second strong peak around 1775–1740 cm⁻¹, you have an anhydride.
- A stretch around 1700–1735 cm⁻¹ could be the C=O of an aldehyde, a ketone, or a carboxylic acid. Got evidence of that —OH curve underlying the C—H stretches? Carboxylic acid. Without that, well, aldehydes should have evidence of a C—H stretch on their C=O at 2850–2700 cm⁻¹, a set of two small peaks that might be obscured by the C—H stretches in the region.
- A stretch around 1750–1730 cm⁻¹ could be the C=O of an ester. Check for two strong C—O stretches in the 1280–1050 cm⁻¹ region.
- A stretch around 1700–1630 cm⁻¹ might mean you have an amide. You can confirm this by checking for an N—H stretch back up the spectrum, but it will only show up for primary and secondary amides. Tertiary amides have no hydrogens on their N's.

Region: 1680–1400 cm⁻¹. The C=C stretch can show up here, the aliphatic C=C around 1680–1600 cm⁻¹, and the aromatic C=C around 1600–1400 cm⁻¹. If the C=C has any H's on it, you can confirm this if there's a C=C—H stretch up around 3000 cm⁻¹.

Region: 1500–600 cm⁻¹. The fingerprint region. A lot of the peaks here are due to the characteristics of the entire molecule and, just like fingerprints, they are unique to every molecule. That's not to say that there aren't characteristic functional group bands here. It's just that assigning them can be problematic.

Region: 850–500 cm⁻¹. The absorptions for halides are around here. The C—Cl stretch can show up almost anywhere in this area. The C—Br stretch is usually relegated to the area below 650.

Regions: 2000–1667 and 910–690 cm⁻¹. Aromatic ring substitution patterns. The aromatic C—H out-of-plane bending around 910–690 generates medium-to-strong peaks that can indicate the substitution pattern of an aromatic compound. The 2000–1667 region can also be used in this manner, but the peaks can be much weaker.

Did you notice? I stopped using the cm⁻¹ notation. While you don't do this when you're writing papers, or in your notebook, or even in your prelab exercises, few people add the word "wavenumbers" or the term "reciprocal centimeters" to the values when they are talking about them.

INFRARED SAMPLE PREPARATION

You can prepare samples for IR spectroscopy easily, but you must strictly adhere to one rule:

No water!

In case you didn't get that the first time:

No water!

Ordinarily, you put the sample between two sodium chloride plates. Yes. Common, ordinary *water-soluble* salt plates. Or mix it with **potassium bromide (KBr)**, another water-soluble salt.

So keep it dry, people.

Liquid Samples

1. Make sure the sample is DRY. NO WATER!

2. Put some of the *dry* sample (1–2 drops) on one plate, then cover it with another plate (Fig. 32.11). The sample should spread out to cover the entire plate. You *don't have to press*. If it doesn't cover well, try turning the top plate to spread the sample, or add more sample.

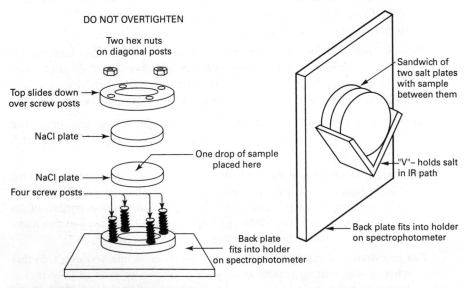

FIGURE 32.11 IR salt plates and holders.

3. Place the sandwich in the IR salt plate holder and cover it with a hold-down plate.

4. Put at least two nuts on the posts of the holder (opposite corners) and spin them down *GENTLY* to hold the plates with an even pressure. *Do not use force!* You'll crack the plates! Remember, these are called salt plate holders and not salt plate smashers.

5. Slide the holder and plate into the bracket on the instrument in the sample beam (closer to you, facing the instrument).

6. Run the spectrum.

7. Take the salt cells out of the instrument and clean them. Often a little acetone followed by drying with a Kimwipe will suffice. Salt plates need a clean, *dry* place to rest in after work. Help them get home.

Since you don't have any other solvents in there, just your liquid compound, you have just prepared a liquid sample, **neat**, meaning *no solvent*. This is the same as a *neat liquid sample,* which is a way of describing any liquid without a solvent in it. It is not to be confused with a "really neat liquid sample," which is a way of expressing your true feelings about your sample.

Solid Samples

The Thin-Film Solid A fast, simple method for getting the IR of a solid. Dissolve your solid in a small amount of a rapidly evaporating solvent—methylene chloride, diethyl ether, pentane—then put a drop of the solution on the plate and wait for the solvent to evaporate. Put the single, solid-coated cell in the salt plate holder, and run the IR as if this were a liquid between salt plates. Some say the salt plate should be gently warmed before dropping the solution on it. Check out P. L. Feist, *J. Chem. Educ.,* **78**, 351 (2001), for all the gory details. Once you've tried this and seen it work, most other methods, including the fairly rapid Nujol mull, get left in the dust.

The Nujol Mull A rapid, inexpensive way to get an IR of solids is to mix them with Nujol, a commercially available mineral oil. Traditionally, this is called "making a Nujol mull," and it is practically idiomatic among chemists. Although you won't see Rexall or Johnson & Johnson mulls, the generic name **mineral oil mull** is often used.

You want to disperse the solid throughout the oil, making the solid transparent enough to IR that the sample will give a usable spectrum. Since mineral oil is a saturated hydrocarbon, it has an IR spectrum all its own. You'll find hydrocarbon bends, stretches, and pushups in the spectrum, but you know where they are, and you ignore them. You can either look at a published reference Nujol spectrum (Fig. 32.12) or run your own if you're not sure where to look.

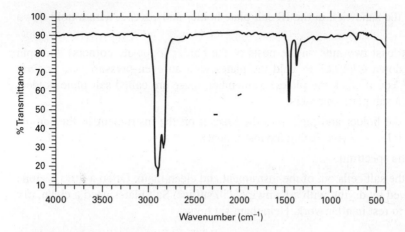

FIGURE 32.12 A published reference Nujol spectrum.

1. Put a small amount of your solid into a tiny agate mortar, and add a few drops of mineral oil.

2. Grind the oil and sample together until the solid is a fine powder *dispersed throughout the oil.*

3. Spread the mull on one salt plate, and cover it with another plate. There should be no air bubbles, just an even film of the solid in the oil.

4. Proceed as if this were a *liquid sample.*

5. Clean the plates with *anhydrous* acetone or ethanol. *NOT WATER!* If you don't have the tiny agate mortar and pestle, try a Witt spot plate and the rounded end of a thick glass rod. The spot plate is a piece of glazed porcelain with dimples in it. Use one as a tiny mortar and the glass rod as a tiny pestle.

And remember to forget the peaks from the Nujol itself.

Solid KBr Methods

KBr methods (hardly ever called potassium bromide methods) consist of making a mixture of your solid (*dry again*) with IR-quality KBr. Regular KBr off the shelf is likely to contain enough nitrate, as KNO_3, to give spurious peaks, so don't use it. After you have opened a container of KBr, dry it and later store it in an oven, with the cap off, at about 110°C to keep the moisture out.

Preparing the Solid Solution

1. At least once in your life, *weigh out* 100 mg of KBr so you'll know how much that is. If you can remember what 100 mg of KBr looks like, you won't have to weigh it out every time you need it for IR.

2. Weigh out 1–2 mg of your *dry, solid* sample. You'll have to weigh out *each sample* because different compounds take up different amounts of space.

3. Pregrind the KBr to a fine powder about the consistency of powdered sugar. Don't take forever, since moisture from the air will be coming in.

Pressing a KBr Disk—the Minipress (Fig. 32.13)

1. Get a clean, dry press and two bolts. Screw one of the bolts about halfway into the press and call that the bottom of the press.

2. Scrape a finely ground mixture of your compound (1–2 mg) and KBr (approximately 100 mg) into the press, so that an even layer covers the bottom bolt.

3. Take the other bolt and turn it in from the top. *Gently* tighten and loosen this bolt at least once to spread the powder evenly on the face of the bottom bolt.

4. Hand-tighten the press again, and then use wrenches to tighten the bolts against each other. Don't use so much force that you turn the heads right off the bolts.

5. Remove both bolts. A KBr pellet containing your sample should be in the press. Transparent is excellent. Translucent will work. If the sample is opaque, you can run the IR, but I don't have much hope of your finding anything.

6. Put the entire press in a holder placed in the analyzing beam of the IR, as in Fig. 32.14. (Don't worry about that yet; I'll get to it in a moment. See "Running the Spectrum.")

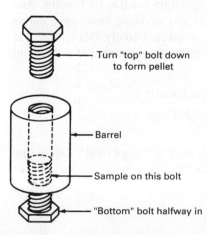

Turn "top" bolt down to form pellet

Barrel

Sample on this bolt

"Bottom" bolt halfway in

FIGURE 32.13 The minipress.

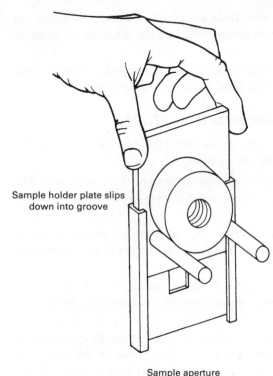

FIGURE 32.14 Putting sample holders with samples into the beam.

Sample holder plate slips down into groove

Sample aperture

RUNNING THE SPECTRUM

Every computer-controlled FTIR seems to have its own proprietary software for getting a spectrum from the optical bench and then for analyzing, marking up, adjusting, and tweaking the collected data into an IR spectrum suitable for framing. And along the way, you might be asked to provide, by way of dialog boxes upon dialog boxes, information about the spectrum you've just taken. Usually, this information gets printed on the spectrum along the side, or in the margins someplace. Some of the things you might be asked for are:

1. *Operator.* The person who ran the spectrum. Usually you.

2. *Sample.* The name of the compound you've just run.

3. *Date.* The day you ran the sample.

4. *Phase.* For KBr, say "solid KBr." A Nujol mull is "Nujol mull." Liquids are either solutions in solvents or "neat liquids," that is, without any solvents, so call them liquids.

5. *Concentration.* For KBr, a solid solution, list milligrams of sample in 100 mg of KBr. For liquids, *neat* is used for liquids without solvents.

6. *Thickness.* Unless you're using solution cells, write **thin film** for **neat liquids**. Leave this blank for KBr samples (unless you've measured the thickness of the KBr pellet, which you shouldn't have done).

7. *Remarks.* Tell where you put your calibration peak, where the sample came from, and anything unusual that someone in another lab might have trouble with when trying to duplicate your work. Don't put this off until the last day of the semester when you can no longer remember the details. Keep a record of the spectrum *in your notebook*.

You now have a perfect IR, suitable for framing and interpreting.

Information about dual-beam IR can be found online at www.wiley.com/college/zubrick.

INTERPRETING IR SPECTRA— FINISHING TOUCHES

IR interpretation—finishing touches can be as simple or as complicated as you'd like to make it. You've already seen how to distinguish alcohols from ketones by **correlation** of the positions and intensities of various peaks in your spectrum with positions listed in **IR tables** or **correlation tables**. This is a fairly standard procedure and is probably covered very well in your textbook. The things that are not in your text are as follows:

1. *Not forgetting the Nujol peaks.* Mineral oil gives huge absorptions from all of the C—H bonds. They'll be the biggest peaks in the spectrum. And every so often, people mistake one of these for something that belongs with the sample.

2. *Nitpicking a spectrum.* Don't try to interpret every wiggle. There is a lot of information in an IR, but sometimes it is confusing. Think about what you're trying to show, and then show it.

3. *Negative results.* Negative results can be very useful. No big peak between 1600 and 2000 cm^{-1}? No carbonyl. Period. No exceptions.

4. *Pigheadedness in interpretation.* Usually a case of "I know what this peak is, so don't confuse me with facts." Infrared is an extremely powerful technique, but there are limitations. You don't have to go hog wild over your IR, though. I know of someone who decided that a small peak was an N—H stretch, and the compound *had* to have nitrogen in it. The facts that the intensity and position of the peak were not quite right and neither a chemical test nor solubility studies indicated nitrogen didn't matter. Oh well.

THE FOURIER TRANSFORM INFRARED (FTIR)

There's even more of a difference in particulars between FTIRs than between dispersive IRs (see online description mentioned earlier) because not only can the basic instrument vary, but it can be attached to any number of different computers of differing configurations running any number of different software programs doing the infrared analysis. Whew. So I'll only highlight the major differences.

The Optical System

The optical system is based on the Michelson interferometer (Fig. 32.15). Infrared energy from the source goes through the sample in a single beam and hits a beam splitter in the interferometer. Half of the light is reflected to a stationary mirror, and the other half passes through the beam splitter to a moving mirror. Both mirrors reflect the light back to the beam splitter, where they recombine to form an interference pattern of constructive and destructive interferences known as an **interferogram**. (Consider the pattern produced when sets of waves from two stones thrown into a pond interact: The peak of one and the peak of the other reinforce each other, giving a bigger peak; the peak of one and the trough of the other can completely destroy each other and produce no displacement.) That interference pattern varies with the displacement of the moving mirror, and this pattern of variation is sent along to the detector.

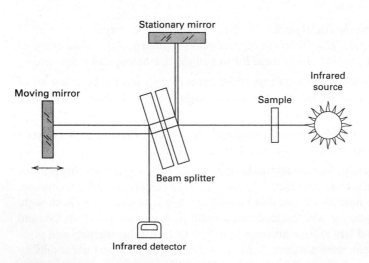

FIGURE 32.15 Michelson interferometer optical system for FTIR.

A computer programmed with the algorithm of the Fourier transformation converts the measured-intensity-versus-mirror-displacement signal (the interferogram) into a plot of intensity versus frequency—our friendly infrared spectrum.

FTIR has a few advantages:

1. *Fellgett's advantage (multiplex advantage).* The interferometer doesn't separate light into individual frequencies like a prism or grating, so every point in the interferogram (spectrum) contains information from each wavelength of infrared radiation from the source.

2. *Jacquinot's advantage (throughput advantage).* The dispersive instrument, with prism or grating, needs slits and other optics so as much as possible of a single wavelength of energy reaches the detector. Not so the interferometer. The entire energy of the source comes roaring through the optical system of the interferometer to the detector, increasing the signal-to-noise ratio of the spectrum.

3. *Conne's advantage (frequency precision).* The dispersive instrument depends on calibration (polystyrene at 1601 cm^{-1}) and the ability of gears and levers to move slits and gratings reproducibly. The FTIR carries its own internal frequency standard, usually a He–Ne gas laser, which serves as the master timing clock and tracks mirror movement and frequency calibration to a precision and accuracy of better than 0.01 wavenumbers (cm^{-1}).

But what about your advantages and your disadvantages?

1. *FTIR is a single-beam instrument.* So you must collect a background spectrum (Fig. 32.16) before you do your sample. The computer program will subtract the background from the background and sample and produce the usual IR spectrum. What's in a background spectrum? Everything that's not the sample. For a salt plate spectrum (Fig. 32.11), clean salt plates in the holder are the background. And for a solution-cell spectrum, the cell with pure solvent is the background. Why? For many cells and holder combinations, the aperture is smaller, or the salt plates absorb a bit, or the solvent absorbs a lot, and these must be compensated for.

2. *The FTIR is fast.* I can get the spectrum of a compound on my computer screen in about 0.21 minute, considerably faster than any dispersive instrument.

3. *The FTIR is accurate.* So you don't have to run a reference polystyrene spectrum. Unfortunately, we plot the spectra on inexpensive copier paper, and if you don't mark relevant peaks on the screen with the software, you may never get those really accurate frequencies to 0.01 cm^{-1}.

4. *The FTIR computer programs automatically scale the spectrum.* So you don't have to set the zero and 100%. Unfortunately, the scaling is mindless, and you can be fooled if you're not careful. Figure 32.17 could possibly be an organic acid. See the broad hydrogen-bonded OH peak at about 3400 cm^{-1}? (The consequences of not marking the exact frequency with the computer program

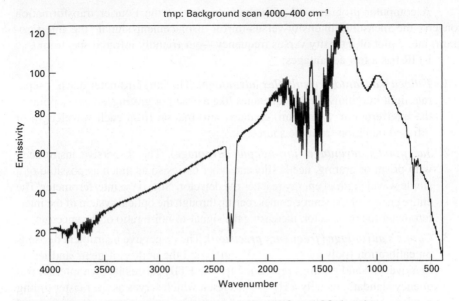

FIGURE 32.16 FTIR background spectrum. Note H_2O and CO_2 bands.

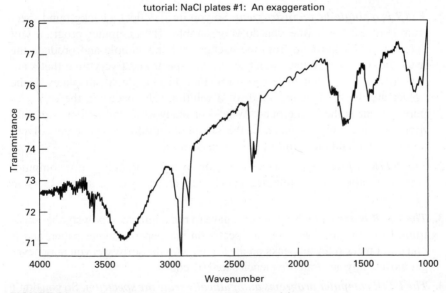

FIGURE 32.17 Salt plate spectrum transmittance trick.

should be evident now.) But the C=O at 1750 cm⁻¹ isn't as strong as it should be and . . . Fooled ya! Look at the transmittance scale on the left-hand side. It only goes from 70% to 76%. This is a spectrum of some student-abused salt plates, and these huge peaks and bands are actually tiny wiggles (Fig. 32.18). The computer program mindlessly took the maximum and minimum wiggling in the spectrum and decided to give it to you at full-screen height.

Admittedly, this is an unusual case. Since my background spectrum was taken with an empty salt plate holder in the beam, my salt plates became the spectrum. Had I run the background with the holder and the cells, any bumps from the salt plates would have been canceled. And to get that 0-to-75% transmittance spectrum, I had to go outside the usual operating range of our instrument. Normally, you won't have such problems. But if you are now on your guard, you won't have *these* problems.

5. *The FTIR programs usually have a library spectral search, so you can easily identify your compounds.* Unfortunately, not every sample you'll run is in the database. So you could wind up with a collection of compounds that usually have the same functional groups (and that can be helpful). Close, but no cigar. In that case, you must not succumb to the "I got it off the computer, it must be right" syndrome.

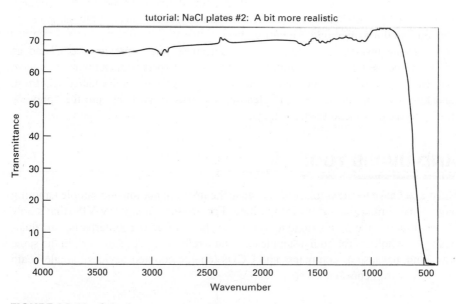

FIGURE 32.18 Salt plate spectrum with some sanity restored.

A REFLECTANCE ATTACHMENT: SOMETHING TO THINK ABOUT

Attenuated Total Reflectance (ATR) lets you get the IR of opaque solids, with little sample preparation. The sample is pressed tightly to a crystal with a high refractive index (Germanium, Zinc Selenide, and so on), and the incoming beam is totally reflected inside the crystal generating an evanescent (that's the word) infrared wave that penetrates a few microns into the solid at every reflection (Fig. 32.19). And unless the sample is something like peanut butter, rather than a flexible polymer, you'll have to put some pressure on the solid to keep it in contact with the crystal. How much? Well, enough to get an infrared spectrum, but not so much you crack the crystal.

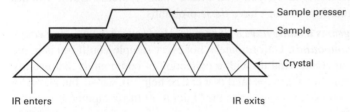

FIGURE 32.19 How an Attenuated Total Reflectance (ATR) crystal works.

More recently, ATR devices use a teeny-tiny diamond as the crystal, mounted below a micrometer. While you can scratch some ATR crystals as your samples get pressed into them, diamond is a bit tougher.

In some instances, you can put the spectrometer into a "preview mode" and, as you press the solid into the crystal, you can watch the spectrum, and when you have a useful result, switch to the "analytical mode" to collect a high-quality spectrum. And after running the sample and cleaning the crystal, go back into the "preview mode" to see how clean your crystal is.

AND UV-VIS TOO!

Only, and I have to stress this, *only* because the instrumentation and sample handling have some similarities is this section here. For instance, many UV-VIS (that's how to speak about them) instruments, just like their infrared counterparts, are dual-beam, have holders for both a sample cell and a reference cell, and work in the same way. And, because of computers and CCDs (charge coupled devices), single-beam UV-VIS instruments also exist.

ELECTRONS GET TO JUMP

In IR spectrometry, it's the bending and stretching of the bonds between atoms that give rise to the spectrum. In UV-VIS, outer electrons are excited directly, and it's these electronic transitions that give rise to the spectrum (Fig. 32.20).

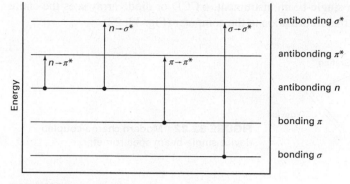

FIGURE 32.20 Possible electronic transitions.

Electrons in π and σ bonds, as well as nonbonding (n) electrons, can be involved.

Most of the UV-VIS spectroscopy of organic compounds involves $n \to \pi^*$ and $\pi \to \pi^*$ transitions that happen to fall in the 200–700-nm range of UV-VIS spectrophotometers.

INSTRUMENT CONFIGURATION

Unlike infra-red instrumentation, ultra violet-visible (UV-VIS) instrumentation has not gone the interferometer/Fourier transform route yet. Apparently, there are no advantages to do so, and a beating in Signal-to-Noise ratio (S/N) relative to dual-beam dispersive instruments, So you're most likely to find a dual-beam UV-Vis instrument (Fig. 32.21) in you lab for a while.

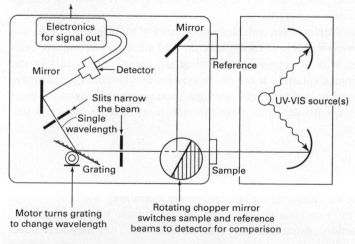

FIGURE 32.21 Dual-beam spectrometer.

In the modern single-beam instrument, a CCD or diode array sees the entire spectrum and sends that information to the computer (Fig. 32. 22).

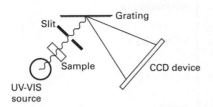

FIGURE 32.22 Modern charge-coupled device single-beam spectrometer.

Source

Traditionally, what amounts to a lightbulb produced the visible spectrum, and a low-pressure hydrogen lamp produced the UV spectrum. Quartz-halogen lamps now may provide a single source for both regions. In all cases, when replacing these lamps, don't leave fingerprints on the glass or quartz.

Sample (and Reference) Cells

VIS-UV cells are square cells 1.0 cm on a side, and can be made of quartz glass, flint glass, or plastic. Only the quartz cells can pass UV as well as visible radiation, so a single set of these would be the most versatile. Plastic cells, while inexpensive, not only don't pass UV, they can dissolve in certain solvents.

Superficially, glass and quartz cells look the same. But if you hold them up to light, and look down the top edges of the cells, quartz cells are bright white, while glass cells tend to be greener. Set up the spectrometer to sit at a UV wavelength (I suggest 254 nm, the same UV wavelength you'd use to light up the green fluorescent substance in TLC plates), and if you get very little or no UV coming through the cells, they're glass.

In a dual-beam spectrometer, consider getting pairs of matched cells. This way you eliminate one more variable in taking spectra. And since they are square cells, they have four sides, with two opposing sides clear and the other two sides frosted (glass, quartz) or rippled (plastic). It can be frustrating, and potentially embarrassing, to find out that you're having trouble getting a good spectrum because the beam is traveling through the frosted sides. Give the cells a one-quarter turn, and start again.

Solvents

Solvents should not be opaque to the regions you're examining with the UV-VIS spectrometer. You can't necessarily use the ubiquitous organic laboratory solvents you've used for reactions, extractions, and recrystallizations. They may have traces of impurities with their own absorptions that can interfere with the absorption spectrum of your sample. You have to use very pure solvents. Because most HPLC units have ultraviolet detectors, HPLC solvents should be considered for your work.

EXERCISES

1. If a bond stretches in the forest, and the dipole moment doesn't change, does it make an absorption?

2. If you prepare a KBr pellet, and you can't see through it, can the IR?

3. A sample gives strong peaks at both 3300 and 1750. What kind of compound might it be? Would it make a difference whether the peak at 3300 is described as very broad or just broad?

4. Someone said that IRs of aromatic compounds are "more spiky" than those of the corresponding aliphatics. Recast this description in less colorful, more boring, but perhaps more accurately detailed terms.

Chapter 33

NUCLEAR MAGNETIC RESONANCE

Nuclear Magnetic Resonance (NMR) takes advantage of the fact that the nuclei of atoms can absorb energy and flip spin states, just like those electrons. And how and where they flip, when plotted, can give a great deal of information about the nature of the compound the nuclei are in, and can lead to the identification of that compound.

The earliest NMR experiments were done on the hydrogen atoms in compounds, and even today if someone says "Show me your NMR," or "Run an NMR on your sample," they are talking about the instrumentation and obtaining the NMR spectrum of the hydrogens in your compound, even though other nuclei can be analyzed this way. Carbon-13 can also spin-flip, and can be analyzed by NMR, but it's always called "C-13 NMR." Phosphorous-31 is "P-31 NMR," Fluorine-19 is "F-19 NMR," and so on. Hydrogens in NMR-speak are almost always called protons, so while you can say "Hydrogen-1 NMR," or even "1H-NMR," understand that "Proton NMR," or "PNMR," all refer to the same thing. And that, at least for the forseeable future, the plain "NMR" will be associated with the proton spectrum.

So after a bit of the theory about the NMR experiment, and bit about **continuous-wave** and **Fourier transform** NMR instrumentation, we'll handle sample preparation, and some basic NMR interpretation (proton, of course).

NUCLEI HAVE SPIN, TOO

Actually, different nuclei can have different spin states: half-integer spin (like those electrons), integer spin, and no spin. The letter I represents these total nuclear spin angular momentum quantum numbers.

If the number of neutrons and the number of protons are both even, the nucleus has no spin. Both the carbon-12 nucleus, with 6 protons and 6 neutrons, and the oxygen-16 nucleus, with 8 protons and 8 neutrons, have zero spin and are NMR-inactive.

If the number of neutrons plus the number of protons is odd, then the nucleus has a half-integer spin quantum number (1/2, 3/2, 5/2, . . .). Protons (hydrogen-1) have a spin of 1/2, and from the selection rule 2I + 1 have two states, +1/2 and −1/2. But there's a catch.

THE MAGNETIC CATCH

Unless an external magnetic field is applied, the two nuclear spin states of protons are said to be **degenerate**. This means that the two states are equal in energy: With an externally applied magnetic field, the energy levels separate, with one state (α) lower in energy, and the other (β) higher in energy. And in the NMR experiment, the point is to supply energy so that the protons can absorb that energy and change, or flip, from the lower to the higher state and then return to the lower state, a process known as **relaxation**.

EVERYBODY LINE UP, FLIP, AND RELAX

So in this externally applied magnetic field, B_0, the protons are distributed between these two energy states, either sort of with the field or sort of against the field, as in Figure 33.1. The spinning nucleus takes on a precessional orbit, much like a spinning top, at a frequency (v) called the Larmor frequency, determined by:

$$v = \gamma B_0$$

where B_0 is the magnetic field strength in tesla, and γ is the gyromagnetic constant (or magnetogyric constant) that's different for each nucleus. For protons $\gamma = 4.26 \times 10^7$ Hz/T, where Hz is the frequency in hertz (formerly cycles per second, with units sec^{-1}) and T is tesla, the magnetic field strength. What's important here is to note that the higher the field strength, the higher the resonant frequency.

Early NMR instruments had a base frequency of 60 megahertz (MHz) using a magnet with a field strength of about 1.4 T. More modern 300-MHz instruments use a 7.05-T magnet to get the protons in resonance. If we sweep either the magnetic field strength or the radio frequency energy so that we go through a proton's goldilocks point (where the magnetic field and the radio frequency are "just right"), the protons will absorb that energy and make the transition to the higher-energy state. Then, as we pass by that goldilocks point, the protons lose that energy by a process called relaxation. In the NMR instrument, there's a coil of wire that detects the energy change involved in this process, and it's that signal that gets amplified and sent to a computer to be recorded.

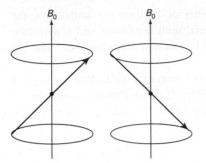

FIGURE 33.1 Processing protons sort of aligned with the external magnetic field in the low-energy and then high-energy state.

A MORE SENSITIVE CENSUS

One of the reasons for using both 60-MHz and 300-MHz resonant (Larmor) frequencies is to show not only that is there a difference in energy between the α and β states, but also that the difference in those frequencies makes a real difference in the NMR.

Since we have the frequencies, we can get the energies by using:

$$E = h\nu$$

For protons at 300 MHz, ΔE is about 0.029 cal/mole; for protons at 60 MHz it's about 0.0058 cal/mole. Now you can get the ratio of the population of protons in the β state, N_β, to that in the α state, N_α, from the Boltzmann equation:

$$\frac{N_b}{N_a} = e^{-\Delta E/kT}$$

Using the ΔE values given, 19.85 cal/K ? mole for the Boltzmann constant, k, and 300 K for the temperature, the ratio is 0.999999 at 60 MHz and 0.999995 at 300 MHz. Not the biggest of differences, but because fewer protons are in the higher-energy state at 300 MHz and 300 K, we get a stronger signal out. Thus, the NMR technique is more sensitive at the higher frequencies.

THE CHEMICAL SHIFT

If all the protons had the same Larmor frequency—resonated at the same frequency—then there would just be one absorption in the NMR for all compounds. Big deal. But that resonant frequency is determined by the magnetic field **at the nucleus**. So, if the magnetic field is different at the nucleus, the goldilocks point will be different for nuclei in different chemical environments, and resonate at different frequencies away from this basic frequency. So when you sweep through a range of Larmor (resonance) frequencies with the instrument, protons in different chemical environments will resonate at these different frequencies and show up at different places in the NMR spectrum.

How small are these differences? Pretty small. The chemical-shift units are in delta (δ), given in parts per million. Parts per million of what? Of the base frequency. So if an NMR signal falls at 1.0 δ, it is at a frequency only 60 Hz from the base in a 60-MHz instrument, or only 300 Hz from the base frequency in a 300-MHz instrument. This should also give you some insight as to why, early on, with only 60-MHz instruments available, NMR spectroscopists would say the chloroform peak is at 432 Hz rather than 7.1 δ. (Actually, they'd have said 432 cps.)

T FOR ONE AND TWO

During the spectrum sweep, protons in different magnetic environments will absorb energy as the sweep goes through the goldilocks point for various protons. When the sweep passes that point for a proton, it can't absorb the exciting energy anymore and starts to lose that energy, to relax. If the protons didn't fall back from their excited states and emit radio-frequency radiation, there would be no energy for the detection coil in the NMR to pick up, and no signal. There are two relaxation processes in NMR:

1. **Spin-lattice relaxation.** The entire NMR sample in the tube is said to be the lattice. All the nuclei in the lattice are in vibrational and rotational motion, making for a very complex magnetic field. Some of the components of the lattice can interact with the protons in the higher-energy state and cause them to lose energy, increasing the rotational and vibrational energy of the entire lattice and resulting in a tiny temperature rise of the entire sample. The time it takes for nuclei to return to the lower-energy state is called T_1.

2. **Spin-spin relaxation.** In this case, a proton in the higher-energy state interacts with a neighboring proton in the lower-energy state, and they exchange quantum states. Thus, the just-excited-and-passed-by proton loses energy as a different proton gains that energy. The time it takes for the proton to lose that energy is called T_2.

BE IT BETTER RESOLVED. . .

Sure, at 300 MHz the energy difference between the α and β is larger than at 60 MHz, and sure, that makes the difference in the population of the two states greater at 300 MHz and smaller at 60 MHz, and sure, that makes the technique more sensitive at 300 MHz than at 60 MHz.

But running at the higher base frequency increases the resolution as well as the sensitivity. As much as we want to get away from using frequencies (Hz) in NMR and instead use δ (parts per million), Hz is always with us. Say there are two NMR signals, one at 3.0 δ and another at 4.0 δ. *They're separated by 1.0 δ no matter what the base frequency of the instrument is.* So, on a 60-MHz instrument they are separated by 60 Hz, and on the 300-MHz instrument their separation is five times greater because there are 300 Hz—not 60 Hz—in the separation. That could make for the difference between reporting an unresolved multiplet and reporting a discernible and analyzable pattern of NMR peaks.

INCREDIBLY BASIC FT-NMR

In the FT version of NMR, you excite all the protons at once, kill the pulse, and record the free induction decay (FID) signal. At that point, your computer takes over and transforms the data from the time domain into the frequency domain. Domain here means x-axis (the domain, remember; y-axis is called the range, eh?), so you eventually get the kind of NMR spectrum you've learned to know and love.

So how do you excite all the protons over the typical range of 10 δ all at once? Lots and lots of teenie NMR transmitters, perhaps? Perhaps it would be best to just blast the sample with a pulse, in which case, if you were paying attention in physics class, or reading articles on the frequency response of audio amplifiers in old issues of *Popular Electronics* instead of going to physics class, you'd know that a square wave can be made up of a series of sine waves of different frequencies. So the pulse itself can excite a range of protons all at once. Then, as the protons relax, the FID is stored away until the next pulse comes along. That FID is also stored away and signal-averaged. When you've collected enough FIDs, the computer then transforms the averaged FID signal into the usual amplitude vs. frequency (in δ parts per million).

NMR SAMPLE PREPARATION

Although FT-NMR has almost taken completely over from continuous wave (CW) instruments, sample preparations for both have some commonalities. Where they differ, I've included breakpoints and explanations for the differences in sample handling. You'll have to pay attention, and not mix them up, OK?

1. *The NMR Sample Tube.* A typical NMR tube is about 180 mm long and 5 mm wide and runs about 5 bucks apiece for the inexpensive model adequate for 60 MHz NMR use. They also have matching color-coded designer caps (Fig. 33.2). Don't pick up and hold the tube by its color-coordinated designer cap. The tube, perhaps with your sample in it, could come out of the cap and fall into the instrument or onto the floor. Hold NMR tubes *just below the cap* by the tube itself.

2. *Preparing the sample.* How you prepare your NMR sample not only depends on whether it is a solid or a liquid, but also whether you have a Fourier Transform (FT) IR continuous wave (CW) instrument. Carefully pick out your situation:
   Continuous wave; liquid sample. This is easy. Get a **disposable pipet** and a little **rubber bulb**, and construct a **narrow medicine dropper**. Use this to transfer your sample to the NMR tube. Don't fill it much higher than about 3–4 cm. Without any solvent, this is called, of course, a **neat sample**.

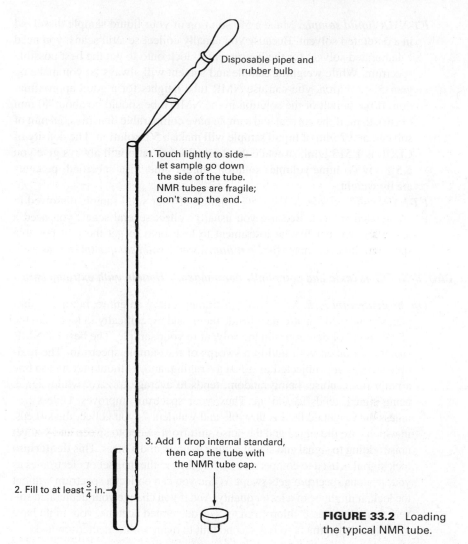

Disposable pipet and
rubber bulb

1. Touch lightly to side—
   let sample go down
   the side of the tube.
   NMR tubes are fragile;
   don't snap the end.

3. Add 1 drop internal standard,
   then cap the tube with
   the NMR tube cap.

2. Fill to at least $\frac{3}{4}$ in.

FIGURE 33.2 Loading
the typical NMR tube.

Continuous wave; solid sample. Get at least 100 mg of your solid sample
ready to put into solution. You'll have to use solvents without any protons,
such as carbon tetrachloride (CCl_4), or **deuterated solvents** so you don't
get this huge distorting response from protons in the solvent. Common deuterated solvents are deuterated chloroform ($CDCl_3$) and deuterated water
(D_2O). If you don't know, can't look up, or aren't told what solvent to
dissolve your solid in, use a few more milligrams and test the solubility in
the protonated versions; the deuterated versions are too expensive for mere
solvent testing. Once you find the right solvent, dissolve your solid, and
then put the solution into your NMR tube.

FT-NMR liquid sample. Make a 5% solution of your liquid sample dissolved in a duterated solvent. Because you usually collect several scans, you need a deuterated solvent for the instrument to lock onto to get the best possible spectrum. While weighing sample and solvent will always let you make up your 5% solution, you can use NMR tube heights for a good approximation. If the height of the solution in the NMR tube should be about 40 mm (4 cm), then, if the solvent and sample have comparable densities, 38 mm of solvent, and 2 mm of liquid sample will make a 5% solution. The density of $CDCl_3$ is 1.513 g/ml, so while the 2 mm/38 mm ratio will always give you a 5% v:v (volume:volume) solution, unless otherwise specified, percents are by weight.

FT-NMR solid sample. Make a 5% solution of your solid sample dissolved in a duterated solvent. Because you usually collect several scans, you need a deuterated solvent for the instrument to lock onto to get the best possible spectrum. Read the notes for *Continuous wave; solid sample* (above) as well.

CAUTION! CCl_4 is toxic and potentially carcinogenic. Handle with extreme care.

On the deuterium lock. Most NMR instruments have an entire, but much simpler, second NMR instrument inside them made specifically to lock onto the NMR signal of deuterium in the solvent in your sample. The best FT-NMR spectra are taken with multiple sweeps of the sample spectrum. The multiple sweeps are subjected to signal averaging, and, without putting too fine a point on it, noise, being random, tends to average to zero, while signal, being signal, tends to add up. Thus, your spectrum improves. Unless . . . unless the magnetic field is unstable and wanders as you collect the sweeps, in which case the signal and the noise shift from sweep to sweep and you get noise adding to signal and signal adding to noise and a mess. This **deuterium lock signal** is used to compensate for any drift in the magnet or electronics as your protons spectrum gets swept. While you can collect a spectrum without the lock, it might be of a lesser quality. And if you change your deuterated solvent from deuterated chloroform to, say, deuterated acetone, you might have to make an adjustment to this second unit to maintain the deuterium lock.

3. *Load the NMR tube.* Use a disposable pipet and rubber bulb to load your NMR tube. Check to make sure the sample is the correct height in the tube, usually somewhere around 4 cm.

4. *Ask about an internal standard.* Usually you use **tetramethlsilane (TMS)** because most other proton signals from any sample you might have fall at *lower frequencies* than that of the protons in TMS, and the TMS signal is set to be 0.0δ (ppm). Sometimes **hexamethyldisiloxane (HMDS)** is used because it doesn't boil out of the NMR tube like TMS can. TMS boils at 26–28°C; HMDS boils at 101°C. Sometimes people add the internal standard to the NMR solvent so all samples using that solvent already have the marker in it, and don't need more. Otherwise, you'd add *only* one—maybe two—drops of internal standard to your sample. Now cap the tube.

5. *Wipe the NMR tube yet again.* You can use a Kimwipe or soft, clean cloth.

6. *Put a spinner on the NMR tube.* Put a plastic air spinner on the tube, and set the height of the spinner from the bottom of the tube in a **depth gauge** on or near the magnet. If the spinner seems too tight or too loose, *immediately stop and get your instructor to check the fit.* If the tube is too loose in the spinner, it may slip in the magnet sample holder and be difficult to remove without breaking something or having to take something apart. If the tube is too tight in the spinner, besides possibly damaging the spinner or breaking the tube, it might not fit the magnet sample holder either and wind up broken or at some weird position in the chamber around the magnet, and might have to be extricated.

   The samples are spun to average out magnetic inequalities that could affect the spectrum.

7. *Load the tube with spinner into the magnet.* Most instruments use an air cushion to load and remove the NMR tube and spinner from the bowels of the magnet. Although there are differences, you start by moving an air valve into the load/unload position, and float the tube with spinner on a cushion of air at the top of the magnet. Then s-l-o-w-l-y moving the valve into the spinner air position, as the air flow shifts from load/unload to spin, the tube with spinner floats down into the magnet. When it comes to rest, it starts spinning, and you take your NMR spectrum.

   Sometimes when the tube and spinner get all the way down into the magnet, the tube refuses to spin. You can move the air valve toward the unload position, raising the tube just a bit, and more slowly lower the tube back to the spinning position. And it can be difficult to look into some magnets to check to see if the tube is spinning or not, so be careful.

   There can also be an air speed needle valve used to control the airflow in the magnet. Do not ever confuse the load/unload valve with the air speed needle valve. Playing with this one will change the spinning speed and if the speed gets too low, the spinning will stop, and if the speed gets too high you'll get vortexing in the tube (at the least) and your spectrum and your reputation will suffer greatly.

8. *Have the instrument make your NMR spectrum.*

9. *Raise (Unload) the tube and spinner from the magnet.* When you're done, you turn the load/unload valve to the unload position, and the tube and spinner float back up out of the magnet where you remove it. *Don't grab the NMR tube by the cap; hold the tube by the glass just under the cap.* You know why.

NMR tubes can be cleaned with a professionally available NMR tube cleaner, or pipe cleaners. Yep, pipe cleaners. Now I seem to recall you could get a roll of this stuff—perhaps that's that twist-tie material people use to tie their tomato plants up with—but if you can't, you can carefully twist together two pipe cleaners at the ends to make one cleaner long enough to do the job. You have to be a little careful putting the twisted section into the tube, but it does work. Leave the tube open to dry . . . and don't even think of blowing unfiltered, dirt- and oil-laden air from the lab air jets into the tube to dry it.

SOME NMR TERMS AND INTERPRETATIONS

I've included a number of NMR spectra to give you some idea of how to read and how to start interpreting NMRs. Obviously, you'll need more than this. See your instructor or any good organic chemistry text for more information.

Unless otherwise specified, the NMR spectra posted here are from the website organized by the National Institute of Advanced Industrial Science and Technology (AIST), Japan (http://riodb01.ibase.aist.go.jp/sdbs/).

The Chemical Shift and TMS Zero

The top spectrum in Figure 33.3 is one of a mixture of chloroform ($CHCl_3$), bromoform ($CHBr_3$), and iodoform (CHI_3), each a **singlet** (single line signal) at 7.256, 6.827, and 4.9 δ (Greek letter delta; ppm, parts per million), respectively. The single proton (hydrogen) on each of these compounds is in a different chemical environments—they have different descriptions. For chloroform ($CHCl_3$), the proton is on a carbon having three chlorines, for bromoform ($CHBr_3$), the proton is on a carbon with three bromines . . . you get the idea. And because chlorine, bromine, and iodine have different electronegativities, the electron cloud surrounding that little proton gets pulled away some in CHI_3, pulled away more in $CHBr_3$, and pulled away the most in $CHCl_3$. So the proton in chloroform has little electron density shielding it from the big bad magnetic field, and is said to be **deshielded**, and appears **downfield** the most.

Remember, the NMR experiment is a goldilocks experiment where everything must be just right to absorb energy and attain resonance, and recall that it has been said that there is some sort of relationship between electricity and magnetism, eh? The electrons in a molecule shield the protons from the external magnetic field, so deshielded, practically nekkid protons need little energy—deshielded; downfield; low energy—to flip, while shielded protons need to be subjected to higher energies—shielded; upfield; higher energy—to flip. Because the three signals, then, are at different places on the spectrum, we say that each proton (or in this case, compound) has a different **chemical shift**.

Now look at the extreme right of the NMR spectrum. That single, sharp peak comes from the protons in the **internal standard, Tetranethylsilane** (TMS). This signal is *defined as zero,* and all other values for the **chemical shift** are taken from this point. Again, the units are parts per million (ppm) and use the Greek letter delta, δ: δ 0.0. Protons of almost all other compounds you'll see will give signals *to the left of zero*: positive values, **shifted downfield** from TMS. [Yes, there are compounds that give NMR signals **shifted upfield** from TMS: negative δ values.] The terms **upfield** and **downfield** are also used as directions relative to where you point your finger on the NMR spectrum: Signals to the *right* of where you are, are **upfield**. Signals to the *left* of where you are are **downfield**.

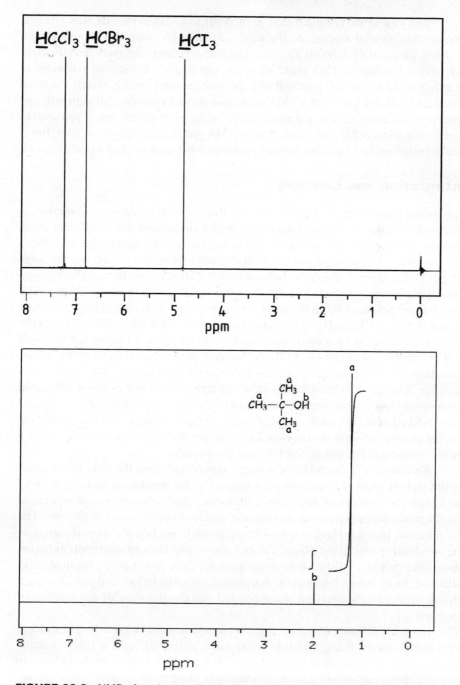

FIGURE 33.3 NMR of a mixture of chloroform, bromoform, and iodoform (top); NMR of t-butyl alcohol spectrum #1 (bottom).

Now look at the chemical shift (δ or ppm) axis one more time. The left side stops a little bit after 8 ppm, and the right side stops just a smidge into negative δ territory. That's so the left side shows a complete digit and hash mark, and the right side shows a complete TMS peak, including the ringing. Setting the computer to exactly 8.0 and exactly 0.0 ppm will still give you the three lines at exactly the same chemical shifts, but your half a TMS peak, and probably missing 8.0 digit will just seem a bit odd on the screen, and possibly a bit odder when printed out. If you want to specifically select or blow up a part of your NMR spectrum, by all means do it. But if you're including the TMS zero in your spectrum, don't print out half a peak. Uncool.

Integration and Labeling

The bottom spectrum in Figure 33.3 is that of *t*-butyl alcohol (2-methyl-2-propanol). Unlike the top picture which was a mixture of three different compounds, this is only one compound, yet it has two singlet signals (peaks). From left-to-right, the tiny singlet comes from the single H on the —OH, and the large singlet comes from all the other protons on the CH_3's. Above the peaks, there are lower-case letters assigned: **b** to the H on the —OH, and **a** to all the others. Why does *b* come before *a*? From the usual drawing of *t*-butyl alcohol having the —OH on the right, the CH_3 —H's get assigned the letter *a*, and the —OH —H gets the letter *b*. Because the O is the electronegative center, then the signal for that —H will be further downfield—that H is deshielded—relative to the H's on the C's of the CH_3's. This happens a lot, as the usual drawings for organic compounds often have the most electronegative element at the right, which makes the NMR signal for protons close to that center be the most to the left. Just be careful.

And all of the H's on the 3 CH_3's are all "a type" because each individual proton has exactly the same description. That "having the same description" can get a little complicated, but will suffice for most compounds.

Because this is the NMR of a single compound, then the ratio of the areas of the signals (peaks) is directly proportional to the number of hydrogens. Way back, spectra were drawn in pen on a uniformly thick chart paper, so you could cut the peaks out, weigh them, and use the masses to get the ratio of protons. The definition of tedious. Have you wondered about those funny curves drawn over the two NMR peaks in the *t*-butyl alcohol spectrum? They're **electronic integrations**, and they can tell you how many protons there are at each chemical shift. Measure the distances between the horizontal lines just before and just after each group, divide by the smallest distance, and viola!—the ratio of the number of protons.

It's a bit difficult, but with my cheap plastic ruler, I get a height ratio of 4.6 mm/0.5 mm or 9.2:1, which is pretty close to the 9:1 ratio in this molecule. Modern computer programs can just spit out the numbers, so you don't have to measure them, but you do have to be careful reading them. Usually the largest—not the smallest—peak is normalized to 1.000, and all other peaks are smaller than that, which is the reverse of what I did with my ruler.

Threaded Interpretations: Spectrum #1 (*t*-butyl alcohol)

For each NMR spectrum, I'll list a few things that might help you interpret your own spectra. Because there might be common elements, I'm not going to repeat them for each one, but just discuss the new features in each, so, to put all of them together, you'll have to follow this thread of interpretation from spectrum to spectrum.

The integration for the *t*-butyl alcohol shows a ratio of 9H to 1H. Nine H's is a lot, and, even better, it's a multiple of three. See 3H, think 1 methyl. See 6H, think 2 methyls. See 9H, think 3 methyls. See hoofprints, think horses. These numbers for protons might not be methyls, but then again, the hoofprints might be zebras, eh? And if **equal** 6H's mean 2 **equal** methyls, consider that the two methyls are both stuck into a single carbon as an isopropyl fragment. If you have the **empirical formula**, you can subtract an extra carbon from the count. The 9H's; 3 methyls can be treated the same way. Consider that the three methyls are all stuck into the same carbon making a *t*-butyl group, and, if you have the empirical formula, you get to subtract four carbons, not just the three, along with 9 H's.

Threaded Interpretations: Spectrum #2 (Toluene) and Spectrum #3 (*p*-Dichlorobenzene) (Fig. 33.4)

First the top figure. NMR: Toluene (5H, 7.2 δ; 3H, 2.34 δ). Well, 3H might be a single methyl, and it is. The other signal is fairly broadly based, centered about 7.2 δ. 7.2! Yikes! The last proton that far down needed 3—count 'em—3 highly electronegative chlorines to pull enough electron density from the proton to deshield it enough to bring it that far down. What's going on here?

The electrons in π bonds interact strongly with the external magnetic field and *generate their own little local magnetic field*. This local field is not *spherically symmetric;* it can shield or deshield protons depending on where the protons are—it's an **anisotropic field**. Figure 33.5 shows some common functional groups—aromatic, carbonyl, alkenyl, and alkynyl groups—with their shielding regions having plusses on them, and their deshielding regions having minuses.

Now take a look at the signal from the 5 protons on the ring at 7.2 δ again. By using the designations a, a', and a", I'm telling you that the descriptions for each of these is NOT the same. Both a's are on carbons close to the CH_3, the a' hydrogens are farther away, and a" is also different. Yet, the SDBS database (and pretty much everyone else) will call all 5 protons on the ring "a-type." It's a shorthand. "Here are all the protons . . . They're all on the same ring . . . They're all the same type . . . a-type." Contrast this with the 7.25 δ signal in the *p*-dichlorobenzene spectrum (#3), where the descriptions of all the protons really are the same, and they, also, are all called "a-type."

Back to the toluene NMR spectrum again. Just as for 3H, 6H, and 9H and methyl groups, you might consider 5H, 10H, and 15H down here about 7.2 δ or so, one, two, and three, singly attached benzene rings called **phenyl groups**. Diphenyl ketone would have an integration amounting to 10 aromatic H's, and triphenylmethanol, with 3 phenyl groups stuck into the carbon, comes out 15H's there.

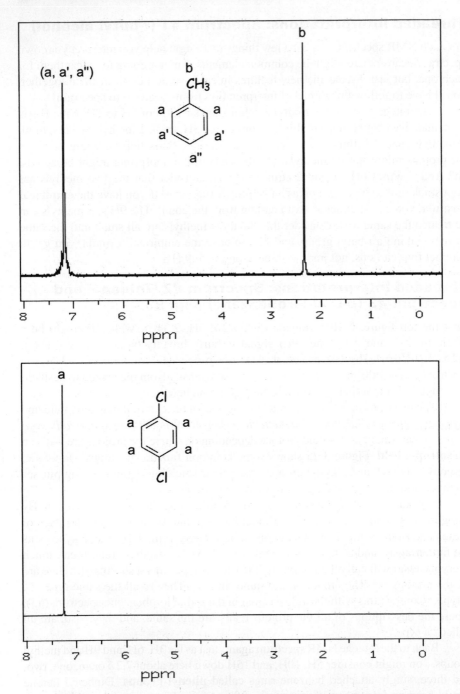

FIGURE 33.4   NMR of toluene spectrum #2 (top); NMR of *p*-dichlorobenzene spectrum #3 (bottom).

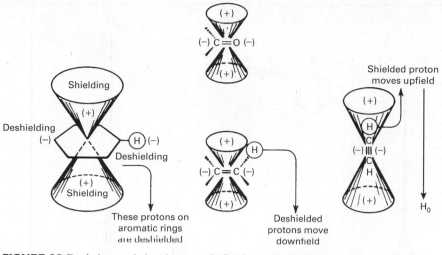

FIGURE 33.5 Anisotropic local magnetic fields on display.

Threaded Interpretations: Spectrum #4 (Ethylbenzene) and Spectrum #5 (A Double Resonance Experiment) (Fig. 33.6)

While the aromatic protons—the "a type"—in ethylbenzene appear usual, the —CH$_2$— and the —CH$_3$ protons are not single lines; they are *split.* **Spin–spin splitting** is the fancy name. Recall protons have a spin of +1/2 or −1/2. If I'm sitting on the methyl group, I can see two protons on the adjacent carbon (—CH$_2$—). (**Adjacent carbon**, remember that.) They spin, so they produce a magnetic field. Which way do they spin? That's the crucial point. Both can spin one way, *plus.* Both can spin one way, *minus.* Or, each can go a different way: one plus, one minus. Over at the methyl group (**adjacent carbon**, right?), you can feel these fields. They add a little, they subtract a little, and they cancel a little. So your methyl group splits into three peaks! It's split by the *two* protons *on the adjacent carbon. Don't confuse this with the fact that there are three protons on the methyl group! That has nothing to DO with it! It is mere coincidence.*

The methyl group shows up as a **triplet** because it is split *by two* protons on the *adjacent carbon.* Now what about the intensities? Why is the middle peak larger? Get out a marker and draw an A on one proton and a B on the other in the —CH$_2$— group (Fig. 33.7). OK. There's only one way for A and B to spin in the same direction: *Both A and B are plus, or both A and B are minus.* But there are *two ways* for them to spin opposite each other: *A plus with B minus; B plus with A minus.* This condition happens *two times.* Both A and B plus happen only *one time.* Both A and B minus happen only one time. So what? So the ratio of the intensities is 1 : 2 : 1. Ha! You got it—a **triplet.** Do this whole business sitting on the —CH$_2$— group. You get a **quartet**— *four lines— because the —CH$_2$— protons are adjacent to a methyl*

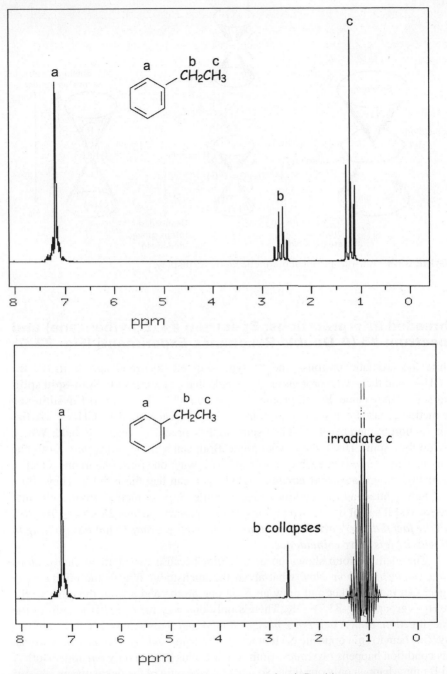

FIGURE 33.6 NMR of ethylbenzene spectrum #4 (top); Double-resonance decoupling in ethylbenzene (bottom).

group. They are split *by THREE to give FOUR lines* (Fig. 33.7). No, this is not all. You can tell that the —CH$_2$— protons and the —CH$_3$ protons split each other by their **coupling constant**, the distance between the split peaks of a single group. Coupling constants are called *J* values and are usually given in hertz (Hz). You used to be able to read this value right from the chart paper, which had a nice grid on it. Perhaps you're lucky. Otherwise, the computer can be set up to produce coupling constants. If you find protons at different chemical shifts and their coupling constants are the same, they're splitting and coupling with each other.

If you think about it, the quartet:triplet pattern is most likely going to be an ethyl group, and ethyl groups are common, so knowing this might be useful. Not only are the *J* values for the quartet and triplet the same, peaks that split or couple each other tend to lean toward each other, as clearly shown in the ethylbenzene NMR. But if you really want to be sure about sets of protons coupling each other, you can do the double resonance experiment.

Remember, it's only a small difference in the populations of two spin states that make the NMR experiment possible, and if the populations were equal, there would be no visible flip. So if you irradiate a set of protons with a strong RF energy field, you equalize that population, stop them from participating in the NMR experiment, and **decouple them from other protons**.

The lower NMR in Fig. 33.6 shows this. A second RF energy transmitter is tuned to the resonance of a particular set of protons, and the set they are coupling with collapses to a single line. This is a reconstruction of what a double resonance

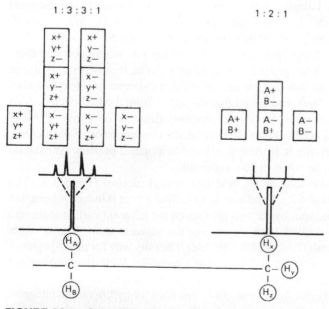

FIGURE 33.7 Spin alignments for the ethyl group.

experiment looks like on a CW instrument, and it shows the zero-beat mixing of the RF energy and the "c protons" as a very large signal. In practice, you just never set the instrument to scan anywhere near the second RF irradiation, and just watch to see what other peaks collapse or change in a way that shows they aren't being split by their neighbors anymore.

Use a Correlation Chart

You'll need a **correlation chart (Fig 33.8)** or **correlation diagram** to help interpret your spectrum, and I'll stick with the ethylbenzene spectrum for a while longer to show how one works. You get the chemical shifts of the signals from your spectrum, and correlate them to a table or chart of functional groups and chemical shifts.

In ethylbenzene, the —CH_3 group is about in the right place (δ 1.23). The δ 7.34 signal is from the aromatic ring, and, sure enough, that's where signals from aromatic rings fall. The δ 2.75 signal from the —CH_2— is a bit trickier to interpret. The chart shows a —CH_3 on a benzene ring in this area. Don't be literal and argue that you don't have a —CH_3, you have a —CH_2 —CH_3. All right, they're different. But the —CH_2— group is on a benzene ring and attached to a —CH_3. That's why those —CH_2— protons are farther downfield; and that's why you don't classify them with ordinary R—CH_2—R protons. Use some sense and judgment.

I've blocked out related groups on the correlation table in Fig. 33.8. Look at the set from δ 3.1 to δ 4.0. They're the area where protons on carbons attached to halogens fall in. Read that again. It's protons on carbons attached to halogens. The more electronegative the halogen on the carbon, the farther downfield the chemical shift of those protons, eh? The electronegative halogen draws electrons to itself, deshielding any protons and making their signals move downfield.

The hydrogen-bonded protons wander all over the lot. Where you find them and how sharp their signals are depend at least on the solvent, the concentration, and the temperature. If you go out and get several different spectra of t-butyl alcohol (one of which is in Fig. 33.3), you'll find that the signal from the hydrogen bonded H varies in chemical shift value. So which laboratory database is "correct?" All of them, of course. You just need to know the solvent, the concentration of the alcohol, and the temperature they ran it at. And if it's that important, perhaps you should record these things for your NMR in your notebook.

Finally, take a close look at the NMR of benzyl alcohol (Fig. 33.9). The hydrogen-bonded —H is at δ 2.33, and sure it looks like a bent rhinoceros horn, but **it's not split**. It should be split by the two protons on the adjacent carbon atom into a triplet. And the CH_2 a singlet? With a proton on the adjacent atom? Those babies should be split into a doublet (one proton adjacent). The only way for the "b protons" and the "c proton" to be singlets is if they aren't coupling. Well, they're singlets; they're decoupled. Hydrogen-bonded hydrogens do not couple, thus they are not split, and they don't split other hydrogens. The same goes for hydrogens on nitrogen. H on O, H on N, and H- on F are hydrogen-bonding hydrogens.

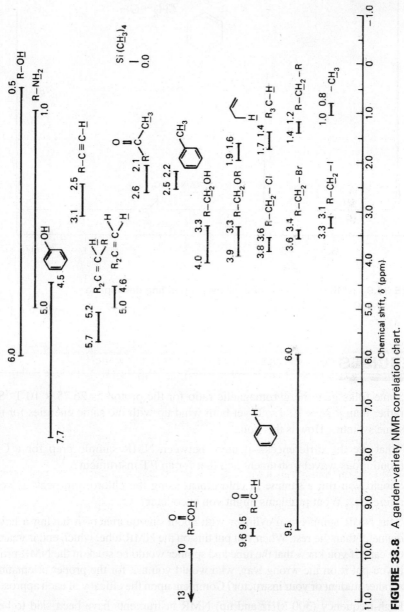

FIGURE 33.8 A garden-variety NMR correlation chart.

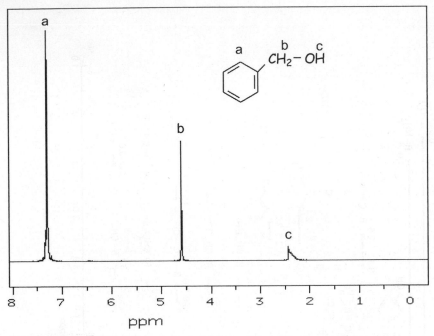

FIGURE 33.9 NMR of benzyl alcohol (shows H-bonding decoupling of protons b and c).

EXERCISES

1. Some folks give the gyromagnetic ratio for the proton as $26.75 \times 10^7 T^{-1}S^{-1}$ rather than 4.26×10^7 Hz/T, yet both wind up with the same energies for the same systems. How is this done?

2. What are the differences—if any—between NMR sample prep for a CW (continuous wave) instrument and that for an FT instrument?

3. Should you run a sample in chloroform using the chloroform peak as your reference?, What problems might you encounter?

4. Your NMR spinner is a cylinder with about one-quarter of it having a larger diameter than the rest. When you put this on the NMR tube, which end is nearest the cap? If you knew that the tube and spinner would be stuck in the NMR probe if you put it on the wrong way, who would you ask for the proper orientation: another student or your instructor? Comment upon the efficacy of each approach.

5. High-frequency (300 MHz and up) NMR instruments have been said to be a magnet, surrounded by liquid helium, surrounded by liquid nitrogen, surrounded by a postdoc. What's a postdoc? Compare the care of a supercon NMR with that of cows on a farm in the context of being able to take a vacation.

INDEX

(Continued)